Thomas D'Aquin Simbe Avore

The institutionalization of modern malaria control

Thomas D'Aquin Simbe Avore

The institutionalization of modern malaria control

Dynamics, resistance, effects: the case of the Centre-Cameroon region(1919-2020)

This book is a translation from the original published under ISBN 978-620-6-70277-1.

Publisher:
Sciencia Scripts
is a trademark of
Dodo Books Indian Ocean Ltd. and OmniScriptum S.R.L publishing group

120 High Road, East Finchley, London, N2 9ED, United Kingdom
Str. Armeneasca 28/1, office 1, Chisinau MD-2012, Republic of Moldova, Europe
Printed at: see last page
ISBN: 978-620-7-74998-0

Contents

I dedicate this book to my thesis director, Mr TCHOUPIE Andre, whose openness, complicity and pertinent observations helped me to see this demanding adventure through to the end.

ACKNOWLEDGEMENTS

This book would not have seen the light of day without the help of a number of people. First and foremost, I would like to thank :

To Dr Charlotte MOUSSI, regional delegate for public health in the Centre, for giving me the opportunity to conduct interviews and collect data within the institution for which she is responsible;

To Doctors Daniel EKOUA, the late Anastasie AKAMBA and Jean Ludovic AMBASSA, respectively directors of the Biyem-Assi, Cite-Verte and Ntui district hospitals, for allowing me to collect data in the hospitals under their care. Without their help, it would not have been possible to analyse the quantitative data contained in this work. Please find here, dear Doctors, the expression of my deep gratitude!

To Mrs Olivia Ngou, Executive Director of the NGOs Malaria No Moore and Impact Sante Afrique, for agreeing to answer my questions during a telephone interview;

To all the people I met during my investigations, both those who took part and those who refused to answer. You have helped me to understand part of the mystery surrounding malaria... I'm fighting with you!

My gratitude!

General introduction

The aim of this book is to study the dynamics of the institutionalization of the fight against malaria in Cameroon. More specifically, it aims to give an account of the process by which the fight against malaria has been gradually and progressively established as a genuine institution, by showing readers the various tribulations and effects associated with this process of institutionalisation. The book is therefore of major social interest, as it enables the political and administrative authorities, health sector stakeholders and local populations to gain a clearer picture of the state of the so-called "modeme" fight against malaria in Cameroon in general, and in the Centre region in particular, by assessing this form of fight. It is also intended to be a support tool for decision-makers and state authorities, enabling them to define better malaria control strategies. To this end, the book sets out proposals for improving the fight against malaria, after reviewing the various obstacles to the effective and efficient implementation of modem public policies to combat the disease.

To speak of a traditional fight against malaria implies that at some point in history there was a traditional fight against the disease. This is understandable, because the peoples of Africa in general, and sub-Saharan Africa in particular, have always had to contend with malaria, whether before, during or after colonisation. But the methods used to combat the disease have not always been the same, depending on the period of history. To speak of a modern fight against malaria also implies that we have disregarded ancient or archaic precepts and methods of response in favour of modern or relatively recent methods. Political anthropologists like Georges BALANDIER believe that what is modern is opposed to what is traditional. Hence the birth of the "tradition and modemity" movement[1] . On the basis of this heuristic postulate, it can be argued that "modern struggle" is opposed to "traditional struggle". If modern malaria control is to be revived as a response to this disease, it must at the same time be acknowledged that this form of control was preceded by another known as "traditional".

Malaria is the most widespread parasitic disease in the world. It constitutes a major risk for more than three billion of the world's population"[2] . From this definition, it can be seen that malaria is an extremely serious disease affecting the well-being of people throughout the world, especially in sub-Saharan Africa. In Cameroon, malaria is still the major endemic disease and the leading cause of illness (morbidity) and death (mortality) in the most vulnerable groups, namely children under five and pregnant women[3] . It would be illusory to think that this disease has not posed a threat to Cameroonian populations today, because before and during the colonial period, they were confronted with the hazards of this

---

[1] BALANDIER (G), <u>Anthropologie politique</u>. Paris, PUF, 1967, pp. 186-217.
[2] LEKE (G.F) et al, <u>Guide pratique de lutte contre le paludisme</u>. Yaounde, L'harmattan, 2010, p. 9.
[3] *Ibid.,* p. 18.

disease, if the work of Robert DEBUSSMANN[4] is anything to go by. In one of his articles, he argues that during the German colonial period, old known diseases such as malaria, intestinal and blood parasites and respiratory diseases continued to occur in Cameroon[5] . This implies that the malaria problem in Cameroon predates the German colonial period. This also implies that methods of combating this disease have always existed, but they had nothing to do with what is done today. They were traditional, because modemitis was not introduced into Africa until the colonial period.

Located in the centre of Cameroon, the Centre region is an emblematic case of this 'tension' between 'tradition' and 'modernity' in the fight against malaria. Indeed, the region is reputed to be the place where French colonisation has had the greatest impact on the indigenous population. Yaounde[6] , its capital, was first one of the capitals of German Cameroon and then the sole capital of French Cameroon. The French y strongly implanted their culture and tradition to such an extent that the nationals of the Centre region were (and still are) considered by the populations of the other regions, particularly those of North Cameroon, as *"little nassara"* or *".* *neo- nassara" (nassara* meaning white here). This labelling is understandable when you consider that the people of Central Cameroon are the Cameroonians who are least attached to their cultural values and who have the greatest propensity to behave like 'whites'. Here, modemity has taken precedence over tradition. From this perspective, the fight against malaria is no exception. Although traditional methods of combating malaria were very important in the Centre region at one point in its history, it has to be said that they were very quickly banished by methods imposed by colonisation. The Centre region is located in the centre of Cameroon and borders five other regions: West, Adamaoua, Littoral, East and South. Elie has ten (10) large departments, thirty (30) health districts, 63 arrondissement communes and one town hall. Its capital is Yaounde, with an estimated population of 3,798,044, an area of 68,926 square kilometres and a population density of 45 inhabitants per km2. Its administrative map is as follows:

---

[4] DEBUSSMANN (R), "Medicalisation et pluralisme au Cameroun allemand : autorite medicale et stratégies profanes", in: Outre-mer, L'Etat et les pratiques administratives en situation coloniale. Tome 90, n° 338-339, 1er semester 2003.
[5] *Ibid.* p. 230.
[6] Originally called "*Jaunde*" (1888-1916), the town we are studying was founded by the German explorer and botanist Georg August Zenker in the German colony of *Kamerun.* After the First World War, in 1919 to be precise, "*Jaunde*" became "Yaounde", renamed by the Iranians after the Germans left Cameroon.

Map 1: Administrative map of the Centre region

It should be pointed out that the term "Centre region" has not always been used, as in the past people spoke of the "Centre province", which itself came from the former "Centre-Sud" province created in 1919 with Yaounde as its capital. The term "Centre province" was used from 1983 to 2008. Presidential decree no. 2008/376 of 12 November 2008 on administrative organisation introduced the name "Centre region". However, for the purposes of this book, the expression "Centre region" is used to designate part of the "Centre-sud" province, the "Centre province" and the "Centre region" itself. It does not refer to an administrative district or a decentralised territorial collectivity, but to a geographical area which has existed since 1919 and which today has thirty (30) health districts, six of which are in Yaounde (Biyem-Assi, Cite-Verte, Djoungolo, Efoulan, Nkolbisson and Nkolndongo) and the rest on the outskirts. Each district is subdivided into health areas. In this way, the Centre region can be seen as an overall system[7] in which there are several health subsystems[8] which operate according to their own logic and coding. Each health sub-system in the Centre region is characterised by a specific habitus[9] , a 'categorisation'[10] [11] and a 'functional differentiation'[m] . What is immediately apparent on observation is that there is a glaring disparity between the

---

[7] By the expression "global system", I mean a territorial entity that is as vast as a State and made up of a multitude of sub-systems without which it would not exist. In more concrete terms, the "global system" here refers to the Centre region understood as an administrative district.

[8] The expression "health subsystem" refers here to the Health District. Used in the plural, it refers to all the Health Districts in the Centre region. As mentioned above, there are 30 of them, including 06 in the city centre and 24 on the outskirts.

[9] This concept comes from Pierre Bourdieu, and is nothing other than *"that* immanent law, *lex insista,* inscribed in the body by identical histories, which is the condition not only for the concertation of practices but also for the practices of concertation". It refers to the culture and societal values of a society, interiorised by all the components of that society. Habitus is what distinguishes one sociological entity from another. Each health sub-system in the Centre region has a habitus that distinguishes it from other sub-systems. For more details see BOURDIEU (P), Le sens pratique, Paris, Minuit, 1980.

[10] This term comes from Sacks Harvey. It refers to the learning engine of our life in society. Each member of a society feels and experiences the world in the same way as his or her fellow members. All this is made possible by the phenomenon of 'categorisation'. Individuals' categorisation of the world varies from one Health District to another in the Centre region.

[11] This is due to the need for specialisation, which has led to the creation of autonomous systems, i.e. systems with their own tasks that are carried out according to their own logic and "coding". See PAPADOPOULOS (Y), Complexite sociale et politiques publiques, Paris, Montchrestien, 1995, p. 18.

urban districts located in the city of Yaounde, in the department of Mfoundi, and the rural districts located in the other departments. This social reality prompted us to focus our attention on this region.

To examine the institutionalisation of the control of malaria in the Centre region is to analyse the process by which this form of control has gradually become established in this geographical area. In other words, it is a question of considering how malaria control has evolved into a veritable institution in the Centre-Cameroon sub-system. It should be pointed out that the first steps towards the institutionalization of malaria control in this sub-system date back to the arrival of the French colonizers in 1916 and especially to the creation of the Centre-Sud province in 1919. As part of the colonial policy of "assimilation", as soon as they entered the region, the French imposed specific methods of combating malaria on the indigenous people through the "colonial military doctors"[12] , whose aim was to banish the traditional methods of combating malaria, which were still being used and were therefore taxed as "barbaric". French colonisation thus became a *'modemising force'* in the Centre region, in other words, a model through which modernisation was universalised[13] . This is why our study will attempt to show how the modern fight against malaria became a veritable institution in the Central Region of Cameroon, within a timeframe ranging from 1919 to 2020. However, this undertaking faces a serious problem of conceptual clarification, as the notion of 'institution' appears in many respects to be one of the most hackneyed concepts in existence[14] .

It was in the 19th century that the notion of institution became, for almost all the founders of sociology, a central concept in their discipline. According to Emile DURKHEIM, "we may (...) call institutions all the beliefs and modes of conduct instituted by the community"[15] . For the French sociologist, institutions are the very object of sociology, because it is through them that social behaviour is established. Jacques ROJOT sees institutions essentially as self-imposed facts that must be constantly taken into account by socio-political actors[16] . Jacques CHEVALLIER, for his part, perceives them as "all the social facts which, over time, have the appearance of an 'objective', 'natural' reality, and are perceived as such by individuals"[17] . Institutions then become "processes by which society organises itself"[18] or, better still, "social phenomena, impersonal and collective, presenting

---

[12] Tasked with implementing "colonial Iranian medicine", these were doctors trained at the Pharo School in Marseille and sent to the Iranian colonies to help domesticate the natives.

[13] BALANDIER (G), <u>Anthropologie politique</u>. *Op. cit...* p. 187.

[14] TCHOUPIE (A), "l'institutionnalisation des deliberations dans 1 espace public au sein des chefferies Bamileke de 1 Ouest-Cameroun", in *Afrique etDeveloppement,* CODESRIA, Vol. XXXIV, 2009, p. 68.

[15] HERMET (G) et al, "Institution/institutionnalisme" in <u>Dictionnaire de la science politique et des institutions politiques</u>. 7ᵉ édition revue et augmentee, Paris, Armand Colin, 2010.

[16] ROJOT (J), *Theorie des organisations,* Paris, Ed. ESKA, 2003, pp. 407-430.

[17] CHEVALLIER (J), <u>Institutions politiques,</u> Pans, LGDJ, 1996, p. 17.

[18] QUERMONNE (J-L), "Les politiques institutionnelles. Essai d'interpretation et de typologie", in GRAWITZ (M) and LECA (J), <u>Traite de science politique</u>, Paris, PUF, tome 4, 1985, p. 62.

permanence, continuity and stability"[19] . Despite their divergences, all these apprehensions of the notion of institution seem to rest on two key elements that serve as their common denominator, namely, deep roots in the social body and permanence over time[20] . It is from this dynamic understanding of the concept of institution that the concept of institutionalisation emerges, which refers to processes and undertakings that tend to organise social models in a stable way[21] . Institutionalisation can be defined as the set of processes that lead to the establishment of one or more institutions. The lexicon of sociology defines it as the process which tends to organise the functioning of society in a stable and lasting way on the basis of regimes[22] . Elijah is born in any situation that extends over time[23] . For this reason, although decision-making procedures will play an important role in this study, the focus will be on the social and historical trajectories by which modeme wrestling has become a veritable institution in the Centre-Cameroon sub-system. This approach will enable us to understand the modal fight against malaria in an institutionalization dynamic. We will consider how modern malaria control gradually and progressively became a belief and a mode of conduct instituted by the community. The central issue underlying this work is therefore the following: how is modern malaria control institutionalised in the Central Cameroon region?

The answer to this question gives rise to the central idea that the modern fight against malaria in the Centre-Cameroon region is being institutionalised through a series of actions which, however, are subject to numerous tribulations and have ambivalent effects. To verify this hypothesis, we need to focus not on 'old institutionalism'[24] or on 'archeo-institutionalism'[25] , but rather on neo-institutionalism. Developed as a reaction against 'old institutionalism' and the formalist-legalist approach that underpinned it[26] , this theoretical framework is far from constituting a unified school of thought. On the contrary, at least three different methods of analysis, all claiming the title of 'neo-institutionalism', have emerged over the last fifteen years: historical neo-institutionalism, sociological neo-institutionalism and rational choice neo-institutionalism[27] . The combination of

---

[19] DURKHEIM quoted by QUERMONNE (J-L), *Idem*, p. 63.

[20] TCHOUPIE (A), "l'institutionnalisation des deliberations dans 1 espace public au sein des chefferies Bamileke de l'Ouest-Cameroun", *Op. cit...* p. 68.

[21] HERMET (G) et al, *ibid.*

[22] DOLLO (C) et al , "Institutionalisation", in *Dictionnaire de sociologie*, 5ᵉ edition, Dalloz, 2017.

[23] BERGER (P) and LUCKMANN (T), La construction sociale de la realite. Pans, Metallic (ed. ong. 1966), 1992, p. 80.

[24] CHEVALLIER (J), Institutions politiques. *Op. cit....* p. 17.

[25] QUANTIN (P), "La decouverte des institutions", in QUANTIN (P), Gouvemer les societes africaines : acteurs et institutions. Paris, Karthala, 2005, p. 13.

[26] During the period of 'old institutionalism', political scientists focused on describing the constitutional and institutional structures of states. This approach was condemned in the 1950s and 1960s for its strong tendency towards description at the expense of explanation. Its formalist-legalist character and its descriptive and a-theoretical nature were roundly criticised. It is for this reason that the prefix 'neo' has been applied to present-day institutionalism. Read LECOURS (A), "L'approche neo-institutionnaliste en science politique : unite ou diversite ?" in Politique et Societe. Vol. 21, 2002, pp. 5-6.

[27] HALL (P.A) and TAYLOR ROSEMARY (C.R), "La science politique et les trois neo-institutionnalismes", in Revue francaise de science politique. 47ᵉ annee, n°3-4, 1997, p. 469.

these three trends provides us with an analytical grid that makes it possible to combine structural interpretation and actors' strategies, contextual definition of agents' policies and choices, and frameworks for predefining their behaviour[28] . Although the historical variance of neo-institutionalism will form the main basis of our analysis, the other two variances (sociological and rational choice) will also be useful for our work.

Historical neo-institutionalism will enable us to highlight the different historical and social trajectories which have contributed to the introduction of modern malaria control in the Centre-Cameroon region. Better still, it will enable us to retrace the historical process by which modern malaria control has gradually imposed itself as a standard of conduct and as the relevant framework for malaria control in this sub-system. Sociological neo-institutionalism will be used to observe the contribution of the cultural variable in the process of introducing modern malaria control. Better still, it will enable us to see how the culture of the local populations attenuates, hinders or 'resists' the hold of modernisation on the traditional values of the populations of the Centre-Cameroon region. Rather than being total, modernisation in the fight against malaria ends up being 'conservative' or 'limited'. As for the neo-institutionalism of rational choice, it will be used to analyse those aspects of the implementation of modern malaria control that are in line with the instrumental behaviours that reveal the calculating instinct of the actors involved in the field of malaria. This state of affairs is a major constraint on the process of institutionalising the fight against malaria.

The mobilisation of all these variants of neo-institutionalism means that, in the end, the fight against malaria has been institutionalised in the Centre region, following a dynamic of gradation and progression, albeit with many tribulations and ambivalent effects. This book is therefore divided into two main sections. The first part is entitled **"The dynamics of institutionalising modern malaria control"**. This part comprises two chapters, the first of which deals with the privileged use of traditional methods up to 2002. It highlights the various actions taken to combat the threat of malaria between 1919 and 2002. These actions are part of a historical process that can be divided into two phases, the first of which runs from 1919 to 1960. This phase was characterised by the privileged intervention of "colonial military doctors" in the field of malaria and by the omnipresence of traditional and prophetic medicine. The second phase, from 1960 to 2002, was characterised by a slight modernisation of the fight against malaria (Chapter I). The second chapter highlights the mechanisms by which the fight against malaria was tightened up from 2002 onwards. Emphasising the 'centrality' of the State in this process of intensification, it analyses the implementation of modern administrative methods and practices in the fight against malaria, which resulted from the accentuation of behaviour-change communication, the increased power of the public authorities

---

[28] TCHOUPIE (A), "l'institutionnalisation des deliberations dans l espace public au sein des chefferies Bamileke de l'Ouest-Cameroun", *Op. cit...* p. 69.

and the free distribution of mosquito nets to households (Chapter II).

The second part is entitled **"The dynamics of institutionalisation: a punctuated equilibrium with ambivalent effects"**. Elie also includes two chapters, the first of which deals with the tribulations of the institutionalization dynamic in the fight against malaria. It analyses the tribulations linked to malaria control institutions and those linked to the field of deployment of social actors that hinder the dynamics of institutionalization of the model malaria control. With regard to the institutions involved in the fight against malaria, the chapter highlights two levels of entanglement, namely 'vertical political entanglement' and 'horizontal political entanglement', which constitute institutional tribulations that punctuate the hardening of the model fight against the malaria problem. With regard to the field of deployment of social actors, the chapter analyses the tribulations linked to the territory, those linked to nationals and those linked to the drug policy they adopt (Chapter III). The second chapter shows that the effects of the institutionalisation of the 'modeme' fight against malaria in the geographical area of Centre-Cameroon are relative, and suggests ways in which this form of 'modeme' fight against malaria can be improved. It shows that, despite changes in the behaviour of nationals and a drop in the number of malaria cases reported in health facilities, the disease still persists in the Centre region. Despite this persistence, however, modern malaria control can still be improved (Chapter IV).

The dynamics of institutionalising modern malaria control

The dynamics of the institutionalization of malaria control means the process by which this form of control has gradually become established and institutionalized. In the geographical area of Centre-Cameroon, this process took place in two distinct phases: 1919-2002 and 2002-2020. Between 1919 and 2002, traditional methods were used predominantly by non-specialists and traditional healers. During the same period, more specifically between 1960 and 2002, the fight against malaria underwent a slight modernisation due to managerial reforms and the rationalisation of traditional medicine. The trajectory of the fight against the disease then underwent a slight bifurcation.

Between 2002 and 2020, the fight against malaria was stepped up with the adoption and implementation of the *"Roll Back Malaria"* programme, which enshrined the principle of "multisectorality"[33] . The trajectory of the fight against malaria then took a sharp bifurcation, leading to rapid institutional changes. The fight against malaria was now subject to a multi-factorial dynamic in the Centre region. Hence its hardening.

Analysing the dynamics of the institutionalization of malaria control in the geographical area of Centre-Cameroon also means, through a historical approach to public policy described as "historical neo-institutionalism"[34] , highlighting the two main phases through which this form of control was gradually established in this sub-system. The first phase ran from 1919 to 2002, and was characterised by privileged recourse to traditional methods of combating malaria **(Chapter 1).** The second phase began in 2002, and was characterised by an intensification of traditional control methods **(Chapter 2).**

---

[33] We'll be back.

[34] Read HALL (P.A) and TAYLOR ROSEMARY (C.R), "La science politique et les trois neo-institutionnalismes", in <u>Revue Francaise de Science Politique</u>. 47ᵉ annee, n°3-4, 1997, pp. 470-476. See also BELAND (D), "Neo-institutionalisme historique et politiques sociales: une perspective sociologique", in <u>Politiques et Societes</u>. vol. 21, 2002.

<u>Chapter 1</u>
Preferential use of
traditional methods
until 2002

It should be pointed out at the outset that Cameroon's history has been dominated by two successive foreign powers, the Germans and the French, at least in the eastern part of the country. The foundations of the modern fight against malaria in Cameroon were laid during this period of history. In fact, the arrival of the Germans in 1884 completely changed the way the people of Cameroon thought about malaria control at that time. The German colonists introduced an "imposition policy" to combat malaria; in other words, they had to abolish the old practices of the so-called "barbarian" peoples, who cured themselves using medicinal plants and placed their fate in the hands of an individual who, according to the Germans, was inexperienced and called a tradipratician or marabout.

In order to sweep away the old malaria-fighting practices that existed at the time, the German administration set up an institution that was hitherto unknown to the local population. This was the hospital, the very first of which was named "Hopital Nachtigal". But faced with this "institutional graft"[35] , the indigenous people did not give in. Hence the ongoing tensions between "professional specialists and independent laymen"[36] . In fact, at the time, there was a certain apathy on the part of the local population, who resisted this German "imposition policy" as best they could. This is why we can affirm with Robert DEBUSSMANN that, during this period of Cameroon's history, "medicalisation came up against a non-formalised institutional barrier"[37] .

The arrival of the French in 1916 marked a turning point. The "policy of imposition" in the fight against malaria was replaced by a "policy of constraint" within the framework of "itinerant offensive medicine". French colonial military medicine accompanied Dr Leon Clovis Eugene JAMOT in his mobile strategy to combat the major endemics. Unlike during the German period, scenes of brutality were commonplace here. Indigenous people were beaten, mistreated and tortured for failing to comply with French "military medicine". Traditional chiefs were condemned, imprisoned and beaten if their administrators did not turn out in large numbers. Mobile teams criss-crossed the villages and anyone who was at home at

---

[35] By this expression, I mean the imposition of Western medicine on the indigenous people by the Germans through the creation of the hospital institution.

[36] See DEBUSSMANN (R), "Medicalisation et pluralisme au Cameroun allemand : autorite medicale et stratégies profanes", In: outre-mer, Tome 90, n° 338-339, 1er semestre 2003, <u>L'Etat et les pratiques administratives en situation coloniale</u>, p. 246.

[37] Ibid. p. 238.

the time of the rally was beaten. It should be pointed out, however, that despite the constraints, the indigenous people resisted French medicine by setting up "bush pharmacies"[38] . It was in the wake of this "policy" of coercion that the geographical area of Central Cameroon came into being in 1919 with the creation of the South-Central Province.

To speak of the privileged use of traditional methods to combat malaria is to analyse the mechanisms and methods used to combat this disease before the adoption and implementation of the *"Roll Back Malaria"* programme[39] by the Cameroonian authorities in 2002. In other words, between 1919 and 2002, malaria control in Central Cameroon was based on unorthodox and more or less archaic practices. During this period, the fight against the malaria problem was subject to specific historical dynamics whose trajectory was marked by bifurcation points and locking mechanisms[40] . Between 1919 and 1960, i.e. during the colonial period, the fight against malaria was carried out through the privileged intervention of "colonial military doctors" and through traditional and prophetic practices **(Section 1)**. Between 1960 and 2002, i.e. during the post-independence period, malaria control was slightly modernised **(Section 2).**

SECTION I

(.'the privileged intervention of "colonial military doctors" and
■ the omnipresence of traditional and prophetic practices
(1919-1960)

Between 1919 and 1960, the fight against malaria in the social health field of Central Cameroon was based on unorthodox practices, hence its archaism and traditional character. It was the result of "colonial medicine" and the practices of the village populations who had settled there. In fact, the French colonial administration there had set up special mechanisms to combat pandemics in general and malaria in particular. These mechanisms, which were of a "new type" for the populations that had settled there long before the arrival of the coloniser, were derived from what was known as "French colonial military medicine", because *"at that time, it was the French military doctors who treated the sick"*[41] . Although these doctors were experts in the medical field, having trained at the Ecole du Pharo in Marseille[42] , they used unorthodox methods to treat patients in the

---

[38] A kind of village pharmacy in which indigenous people treated themselves with medicinal plants.

[39] This is a programme aimed at eradicating malaria in the short term. It was set up in 1998 at the instigation of Dr Gro Harlem Brundland and is a consortium comprising the World Bank, UNDP, UNICEF and WHO. It was approved in Cameroon in 2000 and its committee effectively began operating on 29 July 2002 following decision N°0334/MPS/CAB. For more details, see MOULIOM MOUNGBAKOU (I.B), "Concurrence des therapeutiques traditionnelles et biomedicales dans la lutte contre le paludisme a I'Extreme-nord du Cameroun", in <u>Journal des anthropologues</u>, 2014, pp. 137-157.

[40] THELEN (K), "Comment les institutions evoluent: perspectives d'analyse comparative historique", L'annee de la regulation, 7, 2003-2004, p. 17.

[41] Remarks made by Mr Gottlieb LOBE MONEKOSSO, former Minister of Health in Cameroon, in an interview published in the digital newspaper Cameroon-Info.Net, available online at http://www.cameroon-info.net/article/temoignage-comment-la-fille-du-president-ahidjo-a-ete-admise-au- cuss-131502.

[42] Iranian school where future colonial doctors did their specialist training around the 1920s.

colonies. At the same time, the local populations did not abandon their traditional practices for combating local endemics. In fact, these practices were at one time the most appropriate way of combating malaria.

Thus, between 1919 and 1960, the fight against malaria in the geographical area of Centre-Cameroon was based on the privileged intervention of military doctors, who were non-specialists in the care of patients in the colonies **(paragraph 1). At the** same time, the indigenous people imposed a 'resistance' to this military medicine by emphasising their traditional and ancestral practices for combating disease **(paragraph 2).**

Paragraph 1: The privileged role of colonial military doctors in the field of malaria (1919-1950)

It was between 1919 and 1950 that colonial military doctors became particularly involved in the field of malaria through the implementation of several actions within the framework of *"French colonial military medicine"*. They were both doctors and soldiers, and their mission was to implement "colonial medicine"[43] , which was merely an instrument of the colonisation of Cameroon. At this stage in history, the use of physical force was the most convenient way of ensuring the health of the population and thereby eradicating malaria. The French colonial administration, through its non-specialized military doctors, used coercion in both preventive (A) and curative **(B)** medicine.

A- The privileged role of military doctors in preventive medicine

Preventive medicine refers to 1 set of measures and practices implemented to prevent local populations from falling ill. French colonial military doctors took this form of medicine on board, which is why they initially focused their fight against malaria on the preventive aspect through the strategy known as "pasteurisation"[44] . This was based on a series of restrictive measures whereby non-specialists urged the population to comply strictly with hygiene measures. The strategy took shape in the social arena, with the introduction of a health police force (1) and a health discipline system (2).

1. The role of military doctors in setting up the health police

To prevent cases of malaria, the French administration, following in the footsteps of the German administration, set up a health police force. The aim of military medicine was to encourage the population to respect basic hygiene rules. A "health watch", made up of military personnel, was tasked with raising awareness and then punishing any indigenous people who indulged in acts of insalubrious behaviour. An "informal" form of communication was set up to change behaviour. A number of attitudes were prohibited:

- Eating a midday meal without first taking a bath;

---

[43] Which is "an ambivalent instrument of colonisation, a means of winning hearts and even souls, part of a multifaceted civilising mission". Read DIAW (M), <u>L'appropriation communautaire des cases de sante selon la perspective des populations</u>. These de doctoral en Geographic, Universite Paris Saclay, 2019, p. 22.

[44] Technique borrowed from the French biologist Louis Pasteur, who was the first to discover the means of preventing rabies in 1865.

-   Dispose of household rubbish close to dwellings;
-   Sleeping outdoors at night.

The sanitary police also placed great emphasis on sport as a "sanitary discipline". Domestic arts and nursery corns were given to girls in the early hours of the morning. A roll call was carried out every morning to detect absences from these corns, and if necessary, a penalty was imposed. Equipped with a whip and stationed in front of each concession, the "sanitary policemen" ensured that the indigenes complied strictly with hygiene measures. Sanitary discipline was thus established.

2.   The role of military doctors in establishing health discipline

As a counterpart to the sanitary police, sanitary discipline aimed to promote a type of behaviour that the indigenous people were particularly resistant to:

-   Wash hands with clean water and soap before eating;
-   Take nivaquine every Friday morning before any activity;
-   Clean the latrines every Saturday;
-   Practising sport;
-   Abstain from all traditional rites and self-medication.

To achieve this objective, the military doctors used restrictive measures including "Surveillance", "Control", "Quarantine" and "Isolation" of the local populations[45] . Sanitary discipline also aimed to make the populations of the Centre's social health field the prototype of men that the colonial administration would have liked to have, in accordance with "colonial biopolitics"[46] . In the event of abdication, the "offender" was brought before the disciplinary board, which was held every Saturday at the "Camp d'Ayos" and presided over by a Chief Military Doctor.

In a bid to eradicate the threat of malaria, non-specialists resorted to preventive medicine through a strategy known as "pasteurisation". The aim was to force indigenous people to adopt a certain number of behaviours in order to avoid cases of malaria. As well as prevention, military doctors were also skilled at treating patients as part of curative medicine.

B- The privileged role of military doctors in curative medicine

As the preventive aspect of medicine alone was not enough to combat malaria, military doctors also resorted to curative medicine. This refers to all the measures and practices implemented to treat people who were already ill. The curative aspect of French military medicine used physical force to force people to go to hospital (1) and to treat cases of malaria at home (2).

1. The use of violence in referring malaria patients to hospital

Between 1919 and 1950, the people of the Centre region were obliged to go to hospital if they fell ill at the hands of French military doctors. In fact, the "itinerant offensive medicine" implemented in French East Cameroon left no room for questioning the validity of the hospital institution in the event of malaria. A whole

---

[45] See DEBUSSMANN (R), Op. cit... p. 234.

[46] Neologism used to identify a form of (colonial) power that focuses not on territories but on populations, biopower. Read CHOUMIL (E), "Les ambitions d'un medecin colonial en Afrique :l'histoire du docteur Jean Joseph David", In Classe Internationale, 8 fevrier 2018.

system was set up to force the population to receive the so-called "modern" health care provided by the "Messa dispensary"[47] set up in 1933 (see image 1 below). It is in this sense that David APTER sees colonialism as a 'modemising force', a model through which modernisation has been universalised[48] . According to colonial medicine, the traditional practices of 'barbarian' peoples must be abolished. To achieve this, the sick had to be sought out, treated and cured at all costs[49] . To make this plausible, a team of military doctors was given the task of meticulously exploring the areas to be controlled to their very limits. All the villages were visited, all the inhabitants examined one by one, and a census was taken of each person's name. Those found to be ill were obliged to report to the dispensary the next day for y treatment. In the event of insubordination, physical force was used against the natives, who were described as 'savages'. By reinforcing colonial authority and position, medicalisation was part and parcel of the political and social context in which the doctors operated[50] .

---

[47] Now the Yaounde Central Hospital...

[48] Read BALANDIER (G), <u>Anthropologie politique</u>. Op. cit... p. 187.

[49] See DEBUSSMANN (R), "Medicalisation et pluralisme au Cameroun allemand : autorite medicale et stratégies profanes", Op. cit... p. 232.

[50] Ibid. p. 230.

Source: mystory-iwdical.jimdofree.com/etablissements-cat-l-2/hop-central/

It should be noted that the local population resisted military medicine based on coercion. Elies refused to go to the dispensary and take medicine. Albert Schweitzer described the insubordination of the natives in the following terms:

"They don't want to go to the morning dressing sessions. You have to pick them up one by one. If it's not his turn straight away, he disappears and returns quietly to his home. When the medicine or injection is called, he hears his name, but doesn't move. Even repeated calls don't interest him; he only comes when you take him by the arm"[51] .

To 'break down' the resistance of the indigenous people, the French military doctors worked in synergy with the chiefs of the various villages, who were the 'indigenous health auxiliaries', the real intermediaries between the colonial health services and the people[52] . It was through them that coercion was made more plausible. These chiefs were responsible for imposing penalties on the families of those who refused to go to the dispensary. The most common penalty was a reduction in the food ration. Through this, the "indigenous health auxiliaries" were able to "break" the resistance of the population. In the event of serious and repeated insubordination, the "delinquent" was taken to the "Ayos camp", where the Chief Medical Officer was empowered to discipline the indigenous people, whereas under the colonial regime, the right to punish was reserved for the administrators of the colonies[53] . It therefore follows that the authority of the doctor was hardly established within the walls of the hospital itself[54] . The obligation of the population to stay close to the

---

[51] Quoted by DEBUSSMANN (R), Ibid. pp. 232-233.

[52] Read DIAW (M), L'appropriation communautaire des cases de sante selon la perspective des populations. Op. cit... p. 23.

[53] Read CHOUMIL (E), "Les ambitions d'un medecin colonial en Afrique : l'histoire du docteur Jean Joseph David", In Classe Internationale. 8 february 2018.

[54] See DEBUSSMANN (R), "Medicalisation et pluralisme au Cameroun allemand : autorite medicale et stratégies

hospital was supplemented by rounds and house calls, which made it possible to take care of patients at home by force.

2. The relative systematisation of forced care for patients at home

Between 1919 and 1950, in the social health field of Central Cameroon, at the start of each week, French military doctors would make rounds and visit people's homes as part of the implementation of "vertical malaria control programmes"[55] . The fruit of an institutional pathway[56] mapped out by Dr JAMOT, this strategy enabled non-specialist actors to break down the resistance of populations by taking direct charge of them in their homes, while at the same time intimidating them. Reprimands were issued to those who were prepared to use medicinal plants. The aim of French medicine was to ban self-treatment and cures by local specialists[57] . This is what Doctor Jean Joseph DAVID[58] and his team set out to do in their various tours of Yaounde around 1940. Every home in the Yaounde area was scrupulously inspected with the help of a vehicle bearing a red cross, and physical force was most often used against those who refused to open the door when the doctors arrived. In some respects, the medical rounds were the opposite pole of the hospital. In remote areas where the hospital was inaccessible, the itinerant doctor was able to contact patients[59] . Indigenous people's disobedience in this context was one of the "typical strategies of peasants in the face of power"[60] .

Image 2: A French military doctor in the company of indigenous people during tours and home visits in Yaounde in 1940.

Source: etudescoloniales.canalblog.com

Image 3: Doctor Jean Joseph David in the middle of 5ëance consultation of patients at home (1940).

---

profanes", Op. cit... p. 233.

[55] See <u>Plan Strategique National de Lutte contre le Paludisme au Cameroun 2007-2010</u>. investing in our future, the global fund, to fight AIDS, Tuberculosis and Malaria, p. 16.

[56] See PALIER (B) and SUREL (Y), "Les "trois I" et l'analyse de l'Etat en action", in Revue Franqaise de Science Politique, 2005/1 Vol. 55, p. 7-32.

[57] See DEBUSSMANN (R), Idem, p. 237.

[58] He was an officer in the Troupes Coloniales, graduating in 1929 from the Pharo in Marseille, the school where future colonial doctors did their specialist training. In 1939, he was appointed head of the Haut-Nyong region by Governor Richard Brunot. He treated several malaria patients in Yaounde as part of his tours and house calls.

[59] Read DEBUSSMANN (R), Op. cit... P. 234.

[60] Ibid. p. 233.

Image 4: Treatment /озсёе of malaria patients at home during home visits (1940)

Colonial curative medicine to combat malaria was reflected in the social field by the obligation for the population to go to hospital and by the forced care of patients at home. French military doctors used physical force to make the fight against malaria plausible. This is also why they were described as "non-specialist actors"[61]
. From 1950 onwards, the success of colonial medicine was undeniable and perceptible, thanks to the forced treatment of patients. Almost all endemic diseases had disappeared. The local populations gradually became accustomed to these measures and put aside their old traditional ways of combating malaria. But,

---

[61] See images 2, 3 and 4 in the appendix to this work.

because of the difficulties associated with the rapid and exponential growth of the local population and the ever-increasing demand for health care, the military doctors felt obliged to demand that the indigenous people pay a sum of money in order to receive health care. This led the local population to abandon the hospital in favour of self-medication. It was at this point that traditional and prophetic medicine began to gain ground.

Paragraph 2: The omnipresence of traditional and prophetic medicine (1950-1960)

From 1950 onwards, the fight against malaria in the Centre experienced its first bifurcation due to the rise in power of traditional and prophetic medicine. In fact, from this point onwards, the traditional French medicine was overwhelmed by the strong demand for health care from the local population due to the combination of two factors: the proven effectiveness of French military medicine and the growth of the local population. These two factors forced the French doctors to introduce the payment of a sum of money to receive health care in the dispensaries, which had previously been free. It was at this time that the indigenous people introduced parallel measures to combat malaria. These were traditional medicine (A) and prophetic medicine (B).

A- The omnipresence of traditional medicine

According to the official WHO definition, traditional medicine "refers to health practices, methods, knowledge and beliefs involving the use for medical purposes of plants, animal parts and minerals, spiritual therapies, techniques and manual exercises - separately or in combination - to treat, diagnose and prevent disease or to maintain health"[62] . In industrialised countries, adaptations of certain traditional medical practices are called "complementary", "alternative", "non-conventional" or "parallel", and are the subject of controversy as to their non-scientific nature[63] . The WHO points out that the inappropriate use of traditional medicines or practices can have negative or even dangerous effects[64] .

During the colonial period, despite its prohibition by the French colonial administration, traditional medicine remained firmly rooted in the therapeutic practices of the local populations and even established itself as an alternative to colonial medicine from 1950 onwards. Hence the permanent tensions between "professional specialists and independent laymen"[65] in the geographical area of Central Cameroon. In fact, at this period in history, the practice of traditional medicine was reflected in the strong reliance of indigenous people on local therapies (1) and self-medication (2).

1. Heavy reliance on local therapies

To overcome the limitations of French military medicine, the people living in the social and health field of Central Cameroon turned to local therapies to combat

---

[62] See en.wikipedia.org/wiki/Traditional_Medicine
[63] Ibid.
[64] Ibid.
[65] See DEBUSSMANN (R), Op. cit... p. 246.

malaria around the 1950s. This was the reign of "fundamental traditionalism"[66] as Georges BALANDIER sees it. Natural plants came to the fore as an alternative to Western medicine, given their free nature and proven effectiveness. Malaria treatment in this context was carried out by a "healer" who alone mastered the therapeutic virtues of medicinal plants. To treat the *"Tsid"*[67] , several traditional therapies were used by the healers. The most widespread of these were l'*Efobolo, Miam,* l'*Ekuk,* l'*Esi Ngang,* 1' *Eyem* and 1' *Etombo.* These natural remedies became the most effective in curbing the threat of malaria, so much so that the European doctor was considered only as a last resort when traditional cures had failed[68] . In concrete terms, the treatment process for '*Tsid*' was carried out in a particular way, according to the following dosage:

"Efobolo bark is crushed with the *Miam.* Put the mixture in water and heat over a fire or in the sun. This solution is used to wash the whole body of children suffering from *Tsid,* or fever... After washing, diarrhoea appears. If this treatment is unsuccessful, *Mbubui* leaves are crushed on the tips of the children's fingers. Diarrhoea also followed... The treatment at *Ekuk* seems more serious: the bark is removed by hitting the trunk with a stone. It is left to soak in water in the sun, stirring from time to time with a stick. A cup of this very bitter preparation is drunk in cases of fever or repeated fainting spells... As a maintenance treatment and against *Tsid,* one uses enemas of a decoction of *Esingang* bark, or two tablespoons of the very bitter decoction of *Eyem* bark every day. Infant malaria can be treated with poultices of *Otombo* leaves softened by fire, or with an infusion of its leaves"[69] .

This was the traditional method of treating malaria in the Yaounde region, which existed before the arrival of the Europeans and resurfaced around the 1950s as an alternative to French colonial military medicine: it is the most visible manifestation of "fundamental traditionalism", which attempts to safeguard the values and social and cultural arrangements most closely associated with the past[70] . Natural plants and barks, the virtues of which only healers had mastered, are now the natural response to malaria.

Around the 1950s, local therapies were in high demand for two major reasons. On the one hand, local populations were destitute. In fact, this state of destitution led the sick to be inclined towards traditional medicine because it was free. The traditional practitioner treated the sick free of charge because, according to the *Beti* tradition of the time, he benefited from the grace of his ancestors to be able to cure the sick. On the other hand, emancipation movements were on the rise. Around the 1950s, local populations began to emancipate themselves from the metropolis. Protest movements became commonplace, and recourse to local therapies was seen as a form of 'rebellion' against the French colonial power. In other words, the decolonisation of Cameroon also meant decolonising the fight against endemic

---

[66] A form of traditionalism that attempts to safeguard the values and social and cultural arrangements most endorsed by the past. See BALANDIER (G), <u>Anthropologie politique</u>. Op. cit... p. 203.

[67] In the Beti language, this means "large spleen", which is still synonymous with malaria.

[68] Read DEBUSSMANN (R), Op. cit.... p. 237.

[69] Read COUSTEIX (P-J), "L'art et la pharmacopee des guerisseurs Ewondo (Region de Yaounde)", in <u>Recherches et Etudes Camerounaises</u>. 1961, pp. 48-49.

[70] See BALANDIER (G), <u>Anthropologie politique</u>, Loc. cit... p. 203.

diseases. Hence the strong reliance on local therapies, complemented by the practice of self-medication.

2. Preferential use of self-medication

Self-medication is the use of medication by an individual, either on their own initiative or on the initiative of someone close to them, to treat a condition or symptom that they have identified themselves, without consulting a healthcare professional. Self-medication is most often used when the treatment of a disease is carried out within the family. This is a traditional way of combating pandemics, using medicines of dubious and unconventional origin. Shortly before the independence of French East Cameroon, the local populations of the Centre made extensive use of self-treatment to combat malaria. This method of treatment used natural plants such as *Artemisia, Citronella, Aloevera* and *Eucalyptus.*

As well as going to a traditional doctor in the event of malaria, local people in the Centre's social field also resorted to family treatment. In fact, each family had natural medicinal plants at its disposal to ensure their cure, even though there was no expertise in this field. Here, it was the parents who

played the role of healer and warned their children not to go to the dispensary. Robert DEBUSSMANN gave a perfect illustration of this situation of self-medication, which prevailed during the colonial period: *"it was the parents who were always present, rather than the weakened patients, who took the decision to follow a particular treatment, not to take a particular medicine or to change the dosage, or to leave hospital before the end of the treatment[61]* . Between 1950 and 1960, the relatives of malaria sufferers became the main players in the fight against malaria through self-medication. During the same period of history, prophetic medicine also made its mark in the fight against diseases, particularly malaria.

B- The prevalence of the practice of prophetic medicine

Prophetic medicine was another form of local response to the imposition of French military medicine in the Centre from 1950 onwards. Elie refers to a form of medicine essentially based on the traditions attributed to Mohamet, Jesus Christ, or their ancestors. Elie was characterised by a strong belief in a spiritual guide, whose invocation could resolve a health problem considered here as a mystical attack. In other words, in this context, spiritual power is the key to fighting pandemics.

Prophetic medicine mobilises a set of unorthodox practices that are proscribed in public health matters, such as prayer sessions and the organisation of traditional rites. Before Cameroon's independence, these sociomedical acts were considered to be the most effective treatment for malaria, which was seen as a metasocial attack. The social field of Central Cameroon did not escape this grammar, as prophetic medicine became strongly established there from 1950 onwards. Thus, alongside traditional medicine, the fight against malaria in this sub-system was also carried out through prayer (1) and the organisation of traditional rites (2).

---

[67] Ibid. p. 298.

1. The privileged use of prayer in the treatment of malaria

Prayer is defined as the elevation of the soul towards a divinity to express adoration or veneration, thanks or thanksgiving, or to obtain its graces or favours. Around the 1950s, Christianity was firmly rooted in the geographical area of Central Cameroon and was part of the basic beliefs of the local population. *Yesus-Kristus*[71] was for them an omniscient and omnipotent being endowed with a sumatural power capable of remedying any problem y including those relating to health. It was for this reason that, in the event of malaria, rather than going to a doctor or traditional healer, some people preferred to perform a sequence of prayers invoking the name of *Yesus-Kristus*. Through him, they could find satisfaction.

In practical terms, it was in the parish of Bot-Макак in Nyong et Kelle that a famous Jesuit priest introduced the first spiritual therapies around the 1950s[72] . The priest in question was Father Hebga, who was able to cure malaria sufferers by means of "new-style" cures. He also called on his followers to send their sick children to the church for deliverance in the event of illness[73] [74] . This is how *"people came to see him and confided their sorrows to him. He prayed for them and some found themselves healed'JV* For some people, Father Hebga's prayer had therapeutic virtues that no ordinary doctor could master. Laying hands on the patient's head was the most effective remedy for an episode of malaria. This sociomedical act was seen here as the most effective treatment for malaria, which was seen as a metasocial attack[75] . In other words, for the spiritual cure, malaria is a mystical disease and the answer to it must come from the Lord Jesus. Father Hebga himself said in his day that *"it is not I who heal, but the Lord Jesus, our Saviour"[13]* . The free nature of spiritual therapy led many Central Cameroonians to desert the French colonial hospitals and medical services. Around the 1950s, prayer became the preferred method of fighting malaria. The practice of traditional rites was practically in the same vein.

2. The privileged use of traditional rites in the treatment of malaria

As well as praying, the people of Cameroon's pre-independence Centre region also used traditional rituals to combat malaria effectively, at the same time as circumventing French colonial military medicine. These rites took the form of a sacred cult in which a spiritual guide was invoked. According to *Beti* tradition, this guide had the power to heal. In this context, malaria is perceived as an *"Evu"*, or in the case of the *Pahouin* people[76] [77] , a bad spell or curse that affects the person

---

[71] Jesus Christ in Ewondo.
[72] Read BIRABALUGE Louis, "Apprendre d'un pretre guerisseur : Pere Meinrad-Pierre HEBGA", available online at https://afrique.xaveriens.org/collaborateurs/item/louisbirabaluge-sx
[73] Mrs Messina Tsango, aged 91, told us in an interview on 17 August 2022.
Elie was a devotee of Father Hebga in the 1950s.
[74] BIRABALUGE Louis, Ibidem.
[75] See MOULIOUM MOUNGBAKOU (I.B), "Concurrence des therapeutiques traditionnelles et biomedicales dans la lutte contre le paludisme a I extreme-nord du Cameroun", in Journal des anthropologues, 2014, p. 151.
[76] Quoted by BIRABALUGE Louis, Ibidem.
[77] A group of Bantu-speaking peoples from Cameroon, Equatorial Guinea and Gabon, including the Fang, Beti and Bulu.

whose soul must be purified by a fetishist so that he or she can recover his or her health. The main traditional rites used by the local people to 'solve' any health problem were *Estle, Mevungu* and *Melon. These were* rites of expiation and purification of the soul of the sick person, who was considered to have been struck by a bad spell or curse *(Evu).* These rites were celebrated every Saturday in Yaounde and formed part of the prophetic framework for the fight against malaria (*Tsid).*

As far as the actual treatment is concerned, the fetishists (*Bod Mengan*) carry out a number of mystical practices to rid the patient's soul of the stain that has afflicted it. They know how to help and protect the unfortunate victims of sorcerers. They are the ones who know the rites and prohibitions, and who know how to determine the magical cause of all ills, and therefore how to prevent them[78] . In this context, the sorcerers *(Evu) are* responsible for malaria. This is why, at the start of their consultations, the fetishists sing: *"Us give the evil, I take it away", "Us walk in the dark and I walk in the light", "Us stay behind the hut and I stay in front"*[79] . With these words, the fetishist intends to demonstrate his spiritual and mystical superiority over any evil spirit responsible for the patient's illness. Before resorting to mystical-spiritual therapeutic practices, the healer often demands that patients who refuse treatment be taken to another hut, or even another village, to remove them from the influence of the '*Evu*' that is making them ill[80] . Therapy is then carried out by means of the following mystical-spiritual rites:

- The **"*Esob Nyol*"** is a ritual ablution: the patient goes to a fetishist who makes him recount the events of his life in detail, then sprinkles him with water containing various magical plants and the blood of a goat or chicken[81] ;

- *I.'".-Ilog"* and ***"Mbelekua", which have a*** protective rather than punitive value, and during which the debris of an *"Evu" (Eves Osol)*, taken from the corpse of a sorcerer, is used[79] .

The gratuitousness and effectiveness of these mystical rituals in the fight against diseases, including malaria, undermined French colonial medicine and its cortege of constraints, leading to the desertion of the hospital.

In short, during the colonial period (1919-1960), the fight against malaria in Central Cameroon was archaic and traditional in terms of the methods used to control the disease. These methods gradually became institutionalised and established as the benchmark for malaria control. Between 1919 and 1950, the fight against malaria was the work of the French colonial administration through its non-specialist experts, who used restrictive measures to contain the threat of malaria by means of so-called "itinerant offensive" medicine. This medicine involved the use of physical force to both prevent and treat malaria. This changed with the advent of the emancipation movements orchestrated by Cameroonian nationalists and other

---

[78] See COUSTEIX (P-J), "L'art et la pharmacopee des guerisseurs Ewondo (Region de Yaounde)", Op. cit... p. 18.
Ibid. p. 18.
Ibid. p. 27.
Ibid.

76

77

78

[79] Ibid.

factors such as the growth of the local population. This led to a break with French military medicine. From 1950 onwards, the fight against malaria was based solely on local therapies, other archaic and traditional ways of combating malaria. Traditional and prophetic medicine began to gain ground. Traditional therapies, self-medication, prayers and mystical-traditional rites became relevant in the fight against malaria in the geographical area of Central Cameroon until 1960, the date of the independence of East Cameroon. From this date onwards, there was a slight modernisation of the fight against malaria, which lasted until 2002.

SECTION II

A tegere modernisation of the fight against malaria (1960-2002)

Following Cameroon's independence in 1960, the fight against malaria took on a new dynamic in the Centre-Cameroon sub-system. The trajectory of malaria control, which had been marked out since 1919 by the French colonialists in the context of "itinerant medicine", underwent a real break due to a number of "managerial reforms"[82] . These were not just technical or organisational changes. They are sometimes undertakings of change aimed at deinstitutionalising certain existing institutional arrangements (disqualifying certain regimes, undermining or challenging prevailing values or evaluation criteria, questioning cognitive frameworks taken for granted), reinforcing others, but also legitimising and institutionalising new regulations[83] . In other words, the aim of 'managerial reforms' is both to modify previous bad administrative practices, to retain some of their elements considered to be 'rational', and to bring in new elements that were lacking. In short, 'managerial reforms' aim to rationalise previous administrative practices.

From 1960 onwards, the trajectory of the fight against malaria in the social health field of Centre-Cameroon began to diverge, despite the existing locking mechanisms. This led to a slight modernisation of the fight against malaria. In concrete terms, from 1960 onwards, the dynamics of the institutionalization of malaria control were marked both by 'institutional *layering*' ([84]) and by 'path dependence' ([85]). In other words, from this date onwards, the fight against malaria was professionalised, i.e. it became a matter for professionals with the setting up of new institutions **(paragraph 1).** Despite this process of professionalisation, which was the result of major managerial reforms, traditional and archaic practices persisted, albeit in a slightly more rational form **(paragraph 2).**

---

[82] Read DEMAILLY (L), GIULIANI (F) and MAROY (C), "Le changement institutionnel: processus et acteurs", in Sociologies. Dossiers, Le changement institutionnel, 2019, p. 10.
[83] Ibid., p. 10.
[84] See THELEN (K), "Comment les institutions evoluent: perspectives d'analyse comparative historique", Op. cit... p. 30.
[85] Read PIERSON (P), "Path Dependence, Increasing Returns, and the Study of Politics", in American Political Science Review, 94(2), 2000, pp. 251-267.

Paragraph 1: The beginnings of professionalisation in the fight against malaria

From 1960 onwards, the fight against malaria in the geographical area of Centre-Cameroon was professionalised, i.e. it became a matter for health professionals. This marked a break with the malaria control practices that had existed prior to that date. The celebration of Cameroon's independence put an end to French colonisation, and *a fortiori* to itinerant military medicine. The Cameroonian people's health destiny was now in their own hands. To achieve this, numerous reforms were made to the health system. This led to the introduction of a policy known as "experiments"[86] , which established what was described as a "basic health service". The aim was to provide people with technically sound health care in harmony with their local realities. This could not be plausible without a minimum of reform of the health system in place since the colonial period. It was at this precise moment that institutional changes in the fight against malaria took place, creating "forks in the road" that set historical development on a new course[87] . In other words, the fight against malaria underwent an institutional *layering* ([88]) that led to numerous changes around the 1960s. These changes are the result of a strategy known as '*layering*', which consists of adding new areas to an existing institutional set-up in such a way that they gradually affect the way in which the old areas structure or regulate behaviour[89] .

In concrete terms, in the geographical area of Centre-Cameroon, managerial reforms in the fight against malaria revolved around the professionalisation of the fight against malaria. This professionalisation, which led to non-incremental institutional changes, resulted in the creation of the first reference hospitals (A) and the establishment of the very first medical school and health districts (B).

A- Creation of the first local reference hospitals

Cameroon's post-independence period was marked by major reforms to the health system, aimed at modernising and above all professionalising the fight against malaria in the Central Cameroon region: hence the "institutional sedimentation". This led to the creation of the first local reference hospitals, which laid the foundations for modernising the fight against the disease. Through these hospitals, the medicalisation of malaria gradually became a matter for health professionals, over and above the actions of lay people. There are two types of hospital: the University Hospital Centre (1) and the General Hospital (2).

---

[86] See <u>Plan Strategique National de Lutte contre le Paludisme au Cameroun 2007-2010</u>, investing in our future, the global fund, to fight AIDS, Tuberculosis and Malaria, p. 16.

[87] Read HALL (P.A) and TAYLOR ROSEMARY (C.R), "La science politique et les trois neo-institutionnalisme", in <u>Revue Francaise de Science Politique</u>. 47ᵉ annee, n°3-4, 1997, p. 476.

[88] Institutional sedimentation involves the partial renegotiation of some elements of a given set of institutions, leaving the others unchanged, or the addition of new institutions alongside existing ones, gradually altering the whole institutional edifice. In this case, the innovators bypass the existing institutions and add new segments to them rather than dismantle them. See THELEN (K), "Comment les institutions evoluent perspectives d'analyse comparative historique", Loc. cit... p. 30. See also PALIER (B) and SUREL (Y), "Les "trois I" et l'analyse de l'Etat en action", Op. cit... p. 23.

[89] Read DEMAILLY (L), GIULIANI (F) and MAROY (C), "Le changement institutionnel: processus et acteurs", in <u>Sociologies</u>. Dossiers, Le changement institutionnel, 2019, pp 7-8.

1.   The creation of the Yaounde University Hospital Centre (CHUY)

Following the departure of the French from Cameroon, the very first hospital to be set up in the social health field in Central Cameroon was the Yaounde University Hospital (CHUY)[88] . The creation of this hospital marked a breakthrough in the fight against malaria, as the role of the local doctor became paramount. As a result, French military medicine and traditional, archaic practices disappeared, paving the way for medicine introduced by health professionals.

In practical terms, it was the presidential decree of 28 October 1965 that established the CHUY and institutionalised it as a reference point for patient care in Cameroon, mainly in the Centre region. The CHUY is located in the department of Mfoundi, district of Yaounde VI, in the Melen district. Its main mission is to provide a very high standard of medical and paramedical care. It comprises the following departments

- Department of Internal Medicine and Specialties ;
- Paediatrics and specialities department;
- Multi-purpose and surgical intensive care unit ;
- Hemodialysis unit;
- Gyneco-obstetric department and specialities ;
- Department of Anaesthesiology ;
- Churigie and specialities service;
- Odonto-Stomatology Department;
- ENT department ;
- Laboratory and radiology department ;
- Laundry and sterilisation service.

The CHUY is now a hospital with a multitude of services run by health professionals rather than laymen. It is bringing a new dynamic to the fight against diseases in general and malaria in particular.

Image 5: Main facade of Yaounde University Hospital

2. The creation of the Yaounde General Hospital

The process of modernising the fight against malaria in Central Cameroon led to the creation of the Yaounde General Hospital (HGY) in 1987, in the wake of the creation of the University Hospital. This hospital was financed by the Kingdom of Belgium under a cooperation agreement signed with the Republic of Cameroon. It cost exactly 16 billion CFA francs, took three years to build and was handed over in December 1987. The main mission of the HGY is to provide very high-level medical care piloted by health professionals. Like the CHU, it includes the following departments:

- Internal medicine and specialties department;
- Paediatrics and specialities department;
- Multidisciplinary and surgical intensive care unit;
- Hemodialysis unit ;
- Gyneco-obstetric department and specialities ;
- Department of Anaesthesiology ;
- Churigie and specialities service;
- Odonto-Stomatology Department;
- ENT department ;
- Laboratory and radiology department ;
- Laundry and sterilisation service.

Located in the Ngousso district of Yaounde, the YGH has a capacity of 302 beds and no fewer than ten operating theatres. It is a referral hospital for the people of the Centre region, bringing a new dynamic to the fight against malaria, since the health care provided is provided by health professionals.

Image 6: Main facade of Yaounde General Hospital

B- Creation of the first medical school and health districts

Around 1960, management reforms in Cameroon's health system paved the way for the establishment of other new institutions outside the reference hospitals. At the heart of this socio-institutional dynamic was the fight against malaria, which was spreading at an exponential rate because of 'chemoresistance'. chemoresistance"[90][91] . The aim was therefore to ensure the health of local populations, taking account of their local realities in order to promote "community health". These managerial and institutional reforms led to the creation of the first medical school (1) and the division of the territory into health districts (2).

1. The creation of the University Centre for Health Sciences (CUSS)

The Centre Universitaire des Sciences de la Sante (CUSS) was created in 1969 as part of the drive to professionalise the fight against diseases in general and malaria in particular. The creation of this institution represented a major managerial reform of the Cameroonian health system and a break with French military medicine, which was unsuited to the local realities of the populations of Central Cameroon. Gottlieb Lobe MONEKOSSO, who was the first dean of this establishment, explains the main reason for the creation of the CUSS in these terms: *"At the end of the 1960s, almost ten years after independence, Cameroon felt the need to create a medical faculty with local training, because at that time it was the French military doctors who treated the sick"[99]* . The creation of the CUSS was therefore a response to French colonial medicine, as *"the concern was to adapt teaching to local realities and to focus training on community health"[92]* . Created at the instigation of

---

[90] These are micro-organisms present in the human body that are resistant to all kinds of measures deployed to combat malaria. Chemoresistance is at the root of the spread of malaria mortality and morbidity around 1960 in the Centre region.

[91] Remarks by Mr Gottlieb LOBE MONEKOSSO, former Minister of Health of Cameroon
in an interview published in the digital newspaper Cameroon-Info.Net available online at http://www.cameroon-info.net/article/temoignage-comment-la-fille-du-president-ahidjo-a-ete-admise-au-

cuss-131502.
[91] Ibid.

the WHO, the CUSS was both a genuine health reform and an architectural gem. It can be seen in image 7 below.

Image 7: A view of CUSS, the University Centre for Health Sciences, opened in Yaounde in 1969.

Source: https://www.osidimbea.cm/cameroun-okoba/cameroun-1969/

The CUSS was also set up to promote the professionalisation of the fight against diseases, of which malaria is a major one. Its main mission is to train local doctors to treat patients in a way that takes account of local realities. Conventional medicine gradually became institutionalised, to the detriment of the traditional and archaic practices that had existed in the Centre region before Cameroon's independence. The fight against malaria was henceforth to be led by locally trained doctors. Among the first local doctors trained by the CUSS were personalities such as Maurice NKAM, Fritz NTONE NTONE and Peter NDUMBE. These resource persons and health professionals played a crucial role in the fight against malaria in the geographical area of the Centre by treating patients with modem and conventional therapies.

In addition, the creation of the CUSS was seen as a prelude to the introduction of the "primary health care" policy, based on the slogan *"health for all"*. The aim was to train as many local doctors as possible for deployment in the various basic health services. Given the new epidemiological and malaria situation, health services had to be brought closer to the people. It was in this vein that, in the Centre region around 1990, the health districts were set up to make the fight against malaria more operational on a social level.

2. Setting up health districts

The Cameroonian government's concern to bring health administration closer to local populations, in line with WHO guidelines[93] , led to a major reform of sectoral health policy with the introduction of health districts (HDs) around 1990. In 1992, the Bertoua seminar, which brought together the major players in Cameroon's health system, proposed dividing the national territory into one hundred and twenty

---

[93] See the WHO interregional conference held in Harare in August 1987, which strongly recommended the adoption of the decentralised district health system as a means of achieving the "health for all" objectives by the year 2000.

(120) health districts[94] . However, the directives issued at this meeting came up against difficulties with the demographic criterion. It was the intervention of the World Bank in the context of the sectoral policy applied in urban areas that led to the identification of the first health districts in Cameroon between 1990 and 1993: the Yaounde and Douala health districts. Finally, it was the 1993 National Declaration on the Implementation of the Reorientation of Primary Health Care which imposed legal changes on the public health sector and paved the way for the introduction of the text reorganising the national territory into health districts.

The health district is the operational level for implementing the interventions of the National Health Development Plan (PNDS)[95] . It is made up of health areas including, in particular, health centres built around a base or district hospital. The health area is made up of a village or group of villages with a total population of between 5,000 and 10,000, with a clearly defined territory, and served by a health centre which is responsible for health activities[95] . The aim of the health district is to bring health care closer to the local population and to promote community health in order to combat the problem of malaria effectively.

Thirty (30) health districts were created in the Centre region between 1993 and 1995, six (06) in Yaounde and twenty-four (24) on the outskirts. With the creation of these districts, the fight against malaria becomes more operational, because the health facilities (FOSA) are multiplied and are now close to the local populations. Each of the health districts in the geographical area of Centre-Cameroon has a district hospital (HD), at least one district medical centre (CMA) and integrated health centres (CSI). Malaria control is now carried out by health professionals, most of whom have graduated from the CUSS, and at grassroots level, i.e. in people's homes. In all, there were thirty HDs, seventy CMAs and one thousand one hundred and ninety-nine CSIs in the Centre region around 2000. Tables° 1 and 2 below show the different health districts in the Centre region of Cameroon and their health areas created between 1993 and 1995.

Table n°1

The different health districts of Yaounde and their health areas

Health districts	Health areas
BIYEM-ASSI	Akok Doe, Biscuiterie, Biyem-Assi 1, Biyem-Assi 2, Etoug-Ebe, Melen, Mendong, Mvog Betsi, Nkolbikok, Simbock.
GREEN CITY	Briqueterie, Carriere, Cite Verte,

---

[94] Read OKALLA (R), "Cameroun: de la reorientation des soins de sante primaires au plan national de développement sanitaire", in Bulletin de l'APAD, Un systeme de sante en mutation: le cas du Cameroun. 2001, p. 2.

[95] This is the first operational plan of the sectoral health strategy promoted by the WHO and the FAO, the aim of which is to ensure the health of the population by the year 2000.

[95] Read OKALLA (R), "Cameroun: de la reorientation des soins de sante primaries au plan national de développement sanitaire", Op. cit... p. 6.

	Ekoudou, Messa, Mokolo, Nkolbisson, Nkomkana, Tsinga, Tsinga-Oliga.
DJOUNGOLO	EligEssono, Emana, Essos, Etoa Meki, Mballa II, Mballa V, Mvog Ada, Nfandena, Nkolmesseng, Nkolondom, Nlongkak, Tsinga
	Village.
EFOULAN	Afane Oyoa, Ahala, Efoulan, Ngoaekelle, Nsimeyong, Obili.
NKOLBISSON	Ekorezock, Etetak, Nkolbisson, Nkolnkoumou, Nkolso 0, Nkolso-Oyomabang, Nnom Nnam, Oyom Abang.
NKOLNDONGO	Ekounou, Kondengui, Meyo, Mimboman 1, Minboman 2, Nkolndongo 1, Nkolndongo2, Nkomo, Odza.

Source: Compiled by the author using data from the Institut National de la Statistique.

Table 2
The various peripheral health districts (Centre without Yaounde) and their health areas

Health districts	Health areas
AKONOLINGA	Abem, Akak, Akonolinga urbain, Djoudjoua, Edjom, Ekounou, Emvane Soo, Endom, Mengang, Mengueme Si, Yeme Yeme, Zalom.
AWAE	Awae, Elat-Minkom, Mimbang, Ngat, Nkolessong.
AYOS	Ayos urbain, Efoufoup, Kobdombo, Mang, Mbaka, Mboke, Nganga, Nkoambang, Nyamvoudou, Salla, Yenassa.
BAFIA	Assala, MbamEtlnoub, Bafia 1, Bafia 2, Bafia Rurale, Belamba, Baliama, Bayomen, Bokito, Bongo, Deuk, Donenekeng, Gbwah, Goufan, Kiki, Mouko, Ndikinimeki,
	Ombessa, Roum, Tsekane, Yangben.
EBEBDA	Djounyat, Leka, Ngoksa,

	Nkolebouga.
ELIG-MFOMO	Elig-Mfomo, Nkengue.
ESEKA	Bidjocka, Bondjock, Eseka, Ilanga, Likongue, Макак, Messondo, Mom, Song Badjeck, Song Bayang, Song Mbong.
ESSE	Afanloum, Edzendouan, Esse, Esse Ville, Mveng Essaboutou, Ngondimbele, Ngoungoumou, Ongandi.
EVODOULA	Evodoula, Nkalngaha, Nkolassa, Nloudou.
MBALMAYO	Akoeman, Angonfeme, Assie, Mbalmaya 1, Mbalmayo 2, Mbalmaya Ngallan, Mengueme, Metet, Minlaba, Ngomezap, Nkolmeyang, Nkolya, Olama, Onana Mbessa, Ossoessam, Sep, Zoatoupsi.
MBANDJOCK	Edoumdane, Lembe Yezoum, Mbandjock, Mebolo, Mvebekon, Ndjore, Ndombam, Nkoteng 1, Nyassi, Zoa.
MBANKOMO	Binguela, Ebeba, Mbankomo, Mefomo, Ntouessong, Ntouessong Mbankomo.
MFOU	Essazok, Mfou, Ndangueng, Nkilzok, Nkoabang, Nkongoa, Nsimalen, Omvan.
MONATELE	Evodoula, Eyenmeyong, Monatele, Mvomekak, Ngomo, Nkogbong, Nkolassa, Nkolkosse, Nlongbon, Tala.
NANGA-EBOKO	Bibey, Bissaga, Djassi, Emtse, Mbargue, Metep, Minta, Mvomzock, Nanga Eboko, Ngoulmekong, Nguen, Njombe, Nkoteng, Nsem, Wall.
NDIKIMINEKI	Boutourou, Makenene, Ndikimineki, Ndokwanen,

	Nitoukou, Nyokon.
NGONG-MAPUBI	Bot Макак, Boumyebel, Dibang, Hegba, Mandoumba, Matomb, Mbanda, Mbebe Kikot, Mintaba, Ndongo, Ngog Mapubi, Nguibassal, Ntouleng, Sombo.
NGOUMOU	Akono, Bikok, Bikop, Evindissi, Ngoumou, Nkong Abok, Offoumou Nseleck.
NTUI	Biakoa, Mbangassina, Ndimi, Ndjame, Ngambe Tikar, Ngoro, Nguila, Ntui, Nyamangall, Nyamoko, Talba, Voundou.
OBALA	Batchenga, Edingding, Efock, Ekabita Mendoum, Endinding, Essong, Etoud Ayos, Minkama, Nkol Mekok, Nkolnguem, Nkometou, Obala; Yemesoa.
OKOLA	Ebougsi, Ekekam 3, Elig Yen, Lobo, Mva A, Mvoua, Ngoya, Nkolpoblo, Nlong, Okola, Voa2.
SA'A	Lebamzip, Lepopomo, Ndong Elang, Nkolang, Nkolbogo 1, Nkolmgbana, Nkolmvak, Nlong Onambele, Ondondo, Sa'a.
SOA	Ebang, Koulou, Ngali, Ntoessong, Soa, Ting Melen.
YOKO	Doume, Linte, Makouri, Mankim, Nditam, Ndjole, Ngambe Tikar, Yoko.

Source: Compiled by the author using data from the Institut National de la Statistique.

The delimitation of the territory of the Centre region into districts and health areas represents a major reform of the health system in this sub-system and professionalises the fight against the malaria problem. Through these new institutions, the fight against malaria is now being carried out at grassroots level, i.e. at community level. Most arrondissements have become health districts, with a referral hospital called the District Hospital (HD) and a district hospital called the District Medical Centre (CMA). The various villages in the region become health areas with at least one Integrated Health Centre (IHC). This new territorial network promotes community health and is fully in line with the social objective of "health for all" in the year 2000.

In short, the creation of the first local hospitals, the establishment of the first medical school and the setting up of health districts constitute managerial reforms of the health system and contribute to the process of modernization of the fight against the malaria problem in the geographical area of Central Cameroon. It should nevertheless be noted that, despite this process of modernisation, the traditional malaria control practices that existed before Cameroon's independence persist, albeit in a renovated form. These traditional practices constitute phenomena of institutional feedback and inertia which act as 'blocking mechanisms' to institutional change and to the process of modernisation of the fight against the malaria problem in the geographical area of post-independence Central Cameroon. As a result of their existence, the modernisation of the fight against malaria, which began in the 1960s, became slight or relative.

Paragraph 2: The relative rationalisation of traditional medicine

The period 1960-2002 was characterised by major reforms to the health system in the Centre-Cameroon region. These reforms contributed to the modernisation of malaria control. It should nevertheless be noted that, despite this modernisation process, traditional malaria control practices persist, albeit in a new form: this is the most visible manifestation of "formal traditionalism"[96] as described by Georges BALANDIER. The institutions and social or cultural frameworks of the past (the heritage of the past) are maintained, although their content is modified[97] . It is at this level that the "institutional inertia"[98] lies, holding back the process of institutionalising the modern fight against malaria, and thus bringing about incremental changes. This is in line with the work of historical neo-institutionalism, which is concerned with the effects of past policies. It is based on the idea that past choices set up structures and routines which, as they become institutionalised, restrict the room for manoeuvre of political actors in the present[99] . The concept widely used by public policy specialists to describe this situation is *'path dependence'[100]* . According to Paul PIERSON, this concept suggests that "initial choices of institutional *design* have long-term implications for economic and

---

[96] A form of traditionalism which translates into the maintenance of institutions, social or cultural frameworks whose content has changed; of the past heritage, only certain means are retained - the functions and goals have changed. Read BALANDIER (G), <u>Anthropologie politique</u>. Op. cit... p. 203.

[97] Ibid. p. 203.

[98] See BELAND (D), "Neo-institutionalisme historique et politiques sociales : une perspective sociologique", in <u>Politiques et Societes</u>, vol. 21, 2002, p. 28.

[99] See KUBLER (D) and DE MAILLARD (J), <u>Analyser les politiques publiques</u>, Grenoble, PUG, 2009, p. 137.

[100] This concept comes from economics and is part of a critique of classical theory and some of its presuppositions associated with the idea of general equilibrium and the dynamics of diminishing returns. It was first used by the economist Douglass C. North, for whom the phenomenon of *path dependence* had less to do with technologies themselves than with the behaviour of individuals within institutions. The concept was taken up and inserted into the field of political science by the political scientist Paul Pierson. See PALIER (B) and SUREL (Y), "Les 'trois i' et l'analyse de l'Etat en action", Op. cit... pp. 21-22.

political performance"[101] . For the analysis of public action, it aims to show how past public policies and institutions structure present incentives and resources[102] .

Despite the many institutional reforms of the health system in the geographical area of Centre-Cameroon, the phenomenon of "path *dependence*" can be observed with the relative rationalisation of traditional medicine. This phenomenon is reflected in the 'legalisation' of the practice of traditional medicine (A) and in the creation of community health centres (B) during the post-colonial period.

A- The "legalisation" of traditional medicine

Considered at the beginning of Cameroon's independence by the public authorities as an archaic practice whose therapies had to be banished in order to pave the way for modern and conventional medicine, traditional medicine persisted around the 1980s and became part of the institutional malaria control system through a feedback mechanism. In order to rationalise the fight against malaria, the post-colonial public authorities in the social field of Central Cameroon resorted to the therapeutic practices that had been used during the colonial period, while adapting them to current situations and contingencies: this is the phenomenon of path *dependence*. This phenomenon is reflected in concrete and empirical terms in the ratification of the Alma Ata declaration (1) and the issuing of provisional certificates for the practice of traditional medicine (2).

1. Ratification of the Alma Ata declaration[103]

The Alma Ata Declaration, which took place in 1978 in the USSR, enshrined the "primary health care" approach[104] in the fight against disease. This declaration set out a number of values such as social justice, the right to better health for all, participation and solidarity. It promoted a health system centred on the individual, with a view to the right of everyone to the highest attainable standard of health[105] . By promoting the participation of communities in the fight against disease, the Alma Ata Declaration "resurrects" and encourages the practice of traditional pharmacopeia in African states. Article 7 (7) of this declaration defines what is meant by primary health care at local level and promotes traditional medicine in the following terms:

"primary health care involves both local and referral health care staff - doctors, nurses, midwives, auxiliaries and community workers, as appropriate, as well as **traditional practitioners where necessary** - all of whom are socially and technically prepared to work in teams and meet the health needs expressed by the community".[106]

---

[101] Paul Pierson quoted by PALIER (B) and SUREL (Y), ibid. p. 22.

[102] Ibid. p. 22.

[103] This is a declaration resulting from the Alma Ata conference, which was the first international conference on health care organised by United Nations bodies. This conference took place in Kazakhstan on 12 September 1978 and ended with a declaration known as the Alma Ata declaration in the USSR. The Alma Ata conference was attended by almost 140 countries, including the United States, and took place under the leadership of the charismatic Dr Halfdan Mahler, who read out the final declaration.

[104] This is health care based on practical methods and techniques that are made universally accessible, with the full participation of the beneficiary communities. Communities are becoming major players in the fight against diseases such as malaria.

[105] Read Article VI of the Alma Ata Declaration.

[106] Article VII (7) of the Alma Ata Declaration.

It follows from this article that the Alma Ata Declaration introduces "traditional practitioners" into the institutional circuit for combating pathologies at local level in order to promote primary health care. In 1982, Cameroon ratified this declaration and adopted the implementation of primary health care with the general objective of "bringing all peoples, by the year 2000, to a level of health that will enable them to lead a socially and economically productive life"[107] . Traditional healers thus officially became players in Cameroon's health system, and the fight against malaria was henceforth to be waged with traditional pharmacopeia. Old practices in the fight against malaria, once described as archaic, are resurfacing, have been transposed into the present and are becoming the framework for the fight against this disease: this is the phenomenon of *path dependence.*

In the Centre region, with the implementation of the policy known as "primary health care", traditional medicine is resisting change and is now part of public policies to combat malaria. The starting point was the creation of a department responsible for traditional medicine in the organisational chart of the Ministry of Public Health in the 1980s. This gave the green light to the practice of traditional medicine in the Central Cameroon region. Village communities are now involved in the fight against malaria, with the possibility of organising and setting up village pharmacies. Alongside the "modem" hospitals, there are huts and health villages staffed by traditional practitioners with proven expertise in the field. As a result, the traditional and archaic practices that were corns during the colonial period are being reintroduced in a more rational form in the current policies to combat the threat of malaria effectively. This is precisely where the phenomenon of 'path dependence' comes in, because the political and institutional players of the day had no choice but to call on previous malaria control measures whose effectiveness had been demonstrated in order to respond to a current health need. This is the background to the issuing of provisional certificates for the practice of traditional medicine to certain herbalists.

2. The issuing of provisional certificates for the practice of traditional medicine

Having been considered an archaic and outdated practice in the early 60s, traditional medicine was for a long time sidelined by the public authorities in favour of conventional medicine. With the creation of the CUSS, only health professionals were authorised to provide health care to malaria patients in hospitals. Traditional practitioners were branded as "charlatans", prescribing "grandmother's recipes" to the population. As a result of this stigmatisation, traditional medicine temporarily disappeared from the patient care system.

The ratification of the Alma Ata declaration around the 1980s reintroduced traditional practitioners into the institutional circuit of the response to diseases, particularly malaria, in the Centre region. It was in this spirit that on 22 May 1990, the *"Modem Traditional Clinic International"[109,]* of Dr DEWAH and his team was

---

[107] See <u>Plan Strategique National de Lutte contre le Paludisme au Cameroun 2007-2010</u>. Op. cit... p.

founded.

[108] Elie is located in the Mokolo district at the Hotel Le Progres.

BROS obtains a provisional certificate for the practice of traditional medicine issued by the Minister of Health[108] . With this act, traditional medicine was institutionally recognised as a relevant framework for the fight against malaria in the geographical area of Centre-Cameroon. For Dr DEWAH, white man's medicine had shown its limitations. He said that *"malaria can be beaten with a plant that anyone can have in their garden"*[109] . In the same vein, Dr Moherb's NOKADJI clinic obtained its first provisional certificate for the practice of traditional medicine on 23 May 1990, issued by the Minister of Health[110] . It is clear that traditional medicine is becoming more rational and is adapting to the dynamics of relative modernisation in the fight against malaria. It no longer has anything to do with the archaic form of medicine that existed in the geographical area of Central Cameroon before independence. The issuing of provisional certificates to certain herbalists in 1990 shows that the modern fight against malaria is following the institutional path built by traditional practitioners during the colonial period. The creation of community health centres during the same period further justifies the influence of traditional medicine on current public health policies.

B- Setting up community health centres

The community health centres (CHCs) that sprang up in localities in the Centre-Cameroon region around the 1990s follow the logic of the so-called "abrousse" pharmacies that existed during the colonial period and even much earlier. They are therefore the empirical translation of the phenomenon of path dependence and institutional inertia. In fact, old administrative practices are re-appropriated, albeit in new forms, by the current public authorities in order to better respond to a current public health need. With the introduction of CHCs, traditional and village medicine became more rational and adapted to the dynamic of modernisation and institutional change initiated by the public authorities in 1960.

At local level, alongside the IHCs, we find the CHCs, which aim to promote primary health care with an emphasis on the participation of village communities. In this way, the fight against malaria becomes a matter not only for doctors, but also for the local population, who play an active part in setting up the huts. In practical terms, at local level, the CHC is much more promising in terms of health care than the IHC, given its hybrid nature. In fact, the CHC comprises two poles: one "technocratic", in other words the health pole, and the other "participatory", in other words the community pole[111] . Because it introduces village populations into

---

[108] According to Dr DEWAH in an interview on 27 January 2000, the Minister of Health issued him with a provisional certificate for the practice of traditional medicine in the Centre region.

[109] During an interview with Dr. DEWAH, he told us that he had made these remarks in 1990 at the Carrefour intendance in Yaounde during an awareness campaign attended by a large number of people.

[110] In line with the drive to institutionalise traditional medicine, Dr Moherb's NOKADJI obtained his provisional certificate to practise traditional medicine, issued by the Minister of Health on 23 May 1990. To this end, he opened his first branch in the Mokolo district of Yaounde.

[111] Read DIAW (M), L'appropriation communautaire des cases de sante selon la perspective des populations. These de doctoral en Geographic, Universite Paris Saclay, 2019, p. 37.

the institutional circuit of disease control, CHC 'resuscitates' traditional medicine, which is part of the 'primaries' of village communities.

In the geographical area of Centre-Cameroon, there are six (06) CSCs, all located in rural areas, i.e. areas where the population is strongly attached to traditional practices. These are :

- The Case de Sante Cammunautaire Ines d'Ebolsi in the Awae health district, Mimbang health area;

- The Voundou Community Health Centre in the Ntui health district, Voundou health area;

- The Etam-Niat Community Health Centre in the Ntui health district, Talba health area;

- The Mont-Tama 1 Community Health Centre in the Ntui health district, Talba health area;

- The Maim Douce Ebina 1 Community Health Centre in the Ntui health district, Talba health area;

- The Mbioko II Community Health Centre in the Yoko health district, Ngambe Tikar health area.

The treatment of malaria in these CSCs follows the ancestral logic that prevailed before the arrival of the "whites". ***Efobolo, Esingang and Eyem*** barks, as well as ***Mbubui*** and ***Miam*** leaves, are used by traditional practitioners. The CSCs are in fact a reproduction of the "bush pharmacies" that existed in this area before the arrival of the colonialists. But what is new about the CHC is that it is a hybrid, in the sense that it also contains a doctor who is responsible for treating patients according to the logic of modern medicine. Elie is therefore the palpable embodiment of the fact that the indigenous people "opposed colonial domination, while adhering to certain means of modemity introduced by colonisation"[112] . The modernisation of the fight against malaria in this context is essentially "conservative"[113] .

In short, after Cameroon's independence, the fight against malaria underwent a slight modernisation in the Central Cameroon region. Between 1960 and 2002, the region's health system underwent managerial reform, with new institutions being set up to professionalise the fight against malaria. The first local referral hospitals were created, and are now part of the institutional malaria control system. These were the Yaounde University Hospital and the General Hospital. In addition to these hospitals, there was the first medical school and the health districts. The aim was to modernise and, above all, professionalise the fight against malaria. With the establishment of these institutions, the trajectory of the fight against malaria underwent major changes: this was institutional *layering.*

The modernisation drive that began in 1960 did not ignore the traditional medicine

---

[112] Read BALANDIER (G), <u>Anthropologie politique</u>. Op. cit... pp. 194-195
[113] Read BAYART (J-F), <u>L'Etat en Afrique. La politique du ventre</u>. Paris, Fayart, 1989, p. 157 ff.

that had been established. On the contrary, it rationalised it and adapted it to current situations and contingencies. In other words, traditional medicine persists, albeit in a revamped form, despite the many institutional reforms: this is the phenomenon of path *dependence.* In contrast to the colonial era, post-independence traditional medicine is much more rational and adapts to the dynamics of modernisation. This is how it came to be given a 'legal' character, with the ratification of the Alma Ata declaration and the issuing of provisional certificates for its practice to certain herbalists in Yaounde in 1990. As part of the same drive to modernise, community health centres were set up to promote traditional pharmacopoeia at local level.

In conclusion, between 1919 and 2002, the fight against malaria in the geographical area of Centre-Cameroon was marked by a privileged use of traditional methods. During the colonial period, i.e. from 1919 to 1960, the fight against malaria was characterised by the privileged intervention of non-specialist actors and by the strong presence of traditional and prophetic practices. French military doctors had the privilege of intervening in the field of malaria between 1919 and 1950 and implemented what was known as "offensive itinerant" medicine, both in the field of prevention and treatment. Physical force was used to force local populations to receive modem health care. But with the advent of the emancipation movements around the 1950s, and the exponential growth of the local population, French military medicine was put on the back burner, in favour of traditional and prophetic medicine. At this point, the fight against malaria underwent its first fork in the road since 1919. Local populations were now using local therapies and traditional rites to combat malaria. For the indigenous people, decolonising Cameroon meant decolonising its health system. Traditional and prophetic practices replaced military medicine, which was run by non-specialists, until 1960, when the fight against malaria was slightly modernised.

Between 1960 and 2002, the fight against malaria was slightly modernised as a result of the emergence of health professionals and the relative rationalisation of traditional medicine. The professionalisation of malaria control is reflected in the "managerial reforms" of the health system in Central Cameroon.

These reforms led to the creation of reference hospitals, a large-scale medical school and health districts. Hospitals and doctors became the major players in the institutional system for combating malaria, and the laymen who had previously existed were sidelined. The result is a situation of institutional change that Kathleen THELEN illustrates using the concept of 'institutional *layering*'[114].

Moreover, as part of this process of slight modernisation in the fight against malaria, traditional medicine, once regarded as an archaic method, resurfaced and became relatively rationalised around 1990. With the ratification of the Alma Ata Declaration by the State of Cameroon, and the issuing of provisional certificates for the practice of traditional medicine to certain herbalists, traditional medicine

---

[114] See THELEN (K), "Comment les institutions evoluent: perspectives d'analyse comparative historique", Op. cit... p. 30.

acquired a "legal" character. This mechanism of reproduction and readaptation of previous institutions in the policies of the present corresponds to what Paul PIERSON calls 'path *dependence*'. From this point of view, the processes of institutional development in the fight against malaria that began in 1960 in the Centre-Cameroon region are subject to the constraints imposed by the regions chosen earlier. Hence the "institutional inertia" that perpetuates traditional medicine despite the dynamics of institutional change and modernisation in the fight against malaria. Consequently, rather than being complete, the modernisation of the fight against malaria that began between 1960 and 2002 was relative, slight and limited. We had to wait until 2002 to see a real change in the dynamics of the institutionalization of modern malaria control. In fact, with the adoption of the "*Roll Back* Malaria" programme in 2002, the fight against malaria was stepped up and intensified, putting a definitive end to traditional malaria control methods.

Chapter 2
A tougher approach to modern wrestling from 2002 onwards

From 2002 onwards, the fight against malaria in the Centre-Cameroon region underwent a major change of direction. With the adoption by the United Nations of the Millennium Development Goals (MDGs)[115] [116] and the approval by the Cameroonian authorities of the *"Roll Back Malaria"*™ programme, the fight against malaria was stepped up and intensified in the social and health field of Central Cameroon, leading to "rapid institutional change". It was the effective implementation of the *"Roll Back Malaria"* programme by the Cameroonian authorities on 29 July 2002 that gave new impetus to the fight against malaria in the Centre region. This programme enshrines a new and innovative principle: that of "multisectorality"[117] . In other words, the fight against malaria is now based on a multi-factorial or multi-actor dynamic, i.e. it involves several actors, both public and private, collective and individual, state and non-state.

The involvement of these new and multiple players in the field of malaria, and the instruments and resources they mobilise, are leading to non-incremental institutional change in the malaria control process in the Centre-Cameroon sub-system.

The institutionalisation of the fight against this disease is therefore undergoing a process of "institutional *conversion*" in the sense of Kathleen THELEN[119] . From this perspective, the fight against malaria was oriented towards new mandates, new objectives and new functions. It no longer has anything to do with the practices that prevailed before 2002. In other words, it ceases to be relatively traditional, and becomes truly modern.

The intensification of the fight against malaria from 2002 onwards is the result of the interference of a multitude of actors who were previously outside the process of combating this disease. These new players succeeded in reshaping the original institutional *design* and giving a new direction to the fight against malaria. Within this multi-factorial dynamic, the state occupies a central and predominant position

---

[115] A great deal of attention is devoted to the fight against malaria in Africa.

[116] The "Roll Back Malaria" programme, which aims to eradicate malaria in the short term, was set up in 1998 at the instigation of Dr Gro Harlem Brundtland. It is a consortium comprising the World Bank, UNDP, UNICEF and WHO. It was approved in Cameroon in 2000 and its committee began operating effectively on 29 July 2002 following decision N°0334/MPS/CAB. For more details see MOULIOM MOUNGBAKOU (I.B), "Concurrence des therapeutiques traditionnelles et biomedicales dans le lutte contre le paludisme a I'Extreme-nord du Cameroun", In Journal des anthropologues, 2014, PP. 137-157.

[117] The principle of multisectorality was adopted by UNAIDS in 1996, recommending the integration of a wide range of players in the fight against epidemics, particularly AIDS. This principle has been extended to the fight against pandemics such as malaria since 2000, following the adoption of the "Roll Back Malaria" programme.

in terms of the instruments it mobilises and the actions it carries out in the social field. The other players occupy secondary or peripheral positions. In practical terms, the intensification of the modern fight against malaria is reflected in the intensification of   communication about malaria in order to bring about a change in people's behaviour, and in the increased involvement of the public authorities in the process **(Section 1). Moreover,** the modern fight against malaria has also been stepped up since 2002, with the   organisation of   free   mass   distribution campaigns for mosquito   nets **(Section 2).**

SECTION I

(Emphasis on communication to change

behaviour and increase the power of the public authorities)

The intensification of the fight against the malaria problem from 2002 onwards is reflected in the social field by the intensification of communication on malaria and the increased involvement of the public authorities in the process. Communication on malaria is one of the strategic approaches to the fight against the disease set out in the National Strategic Plan for the Fight against -[119] Read THELEN (K), "Comment les institutions evoluent: perspectives d'analyse comparative historique", Op. cit... p. 32.

Paludisme (PSNLP)[118] focuses on "behaviour change communication" (BCC). This strategic approach is becoming the most practical and effective way of preventing the risk of malaria. The process of implementing this campaign was marked by the strong presence and preponderance of the state, which became the "central" player in the institutional malaria control system, although the role of other players was not negligible.

The rise in power of the public authorities in the field of malaria is reflected in the adoption and implementation of major decisions relating to malaria between 2010 and 2014. In this way, they are building what Jaeho EUN calls the "public rationality"[119] , making malaria a genuine "public problem"[120] . These public authorities, both central and local, are participating in their own way in the process of stepping up the fight against malaria. The following sections will be devoted to an analysis of the increasing importance of behaviour change communication **(Paragraph 1)** and the growing importance of public authorities in the field of malaria **(Paragraph 2).**

Paragraph 1: Emphasis on behaviour change communication

Since the adoption of the *"Roll Back Malaria"* programme in 2002, the population of the Centre-Cameroon area has been the subject of an intense malaria awareness

---

[118] Particularly the 2002-2006 period.

[119] Elie is the basis on which public authorities make decisions, and emerges when public players take on a given social problem. Read EUN (J), <u>Sida et action publique. Une analyse du changement de politiques en France.</u> Paris, L'Harmattan, 2009, p. 171.

[120] That is to say, a problem taken in hand by the public authorities or if it calls on these authorities to solve it. See SHEPPARD (E), "Probleme Public", in BOUSSAGUET (Laurie), JACQUOT (Sophie) and RAVINET (Pauline), Dictionnaire des politiques publiques, 2ᵉ edition, Presses de la fondation nationale des sciences politiques, 2006, p. 354.

43

campaign. This is the result of one of the PSNLP's strategic approaches to combating the disease, based on "Behaviour Change Communication". The aim of this strategy is "to use local communication to encourage all families in Cameroon to put into practice the measures to combat malaria recommended by the Programme".

National Malaria Control Programme (PNLP)"[121] . According to the strategic plan, *"Behaviour Change Communication will aim to induce behaviour within the community that is conducive to reducing malaria-related morbidity and mortality. To support this strategic approach, communication materials will be produced and distributed to the most peripheral levels of the community, and broadcast on the various radio and television channels. Information/debate sessions will be organised in the "health clubs" of the establishments by the NGOs/Associations in each health area. Malaria control days will be commemorated in each health area"[122]* . This means, in other words, that the fight against malaria must be based on a vast campaign to raise people's awareness through communication that is supposed to result in "an endogenous change in the preferences of an interlocutor or an audience"[123] . This brings us back to the work on the behaviourist approach to communication pioneered by Harold D. LASSWELL in 1948. According to this approach, communication is characterised by its effects on the recipient[124] . The aim is to get people in the geographical area of Centre-Cameroon to change their attitudes and behaviour with regard to malaria. It has nothing to do with the peripheral and informal communication campaigns that existed during the colonial period.

In the Centre region, communication on malaria has been stepped up since 2002 as part of the "Behaviour Change Communication" (BCC) campaign. The aim is to promote compliance with hygiene measures and to facilitate the use of impregnated mosquito nets by households. For this reason, in the run-up to the mass distribution of LLINs, the socio-political and institutional players decided to implement a large-scale communication campaign on the use of mosquito nets, known as "Behaviour Change Communication" (BCC). The implementation of this campaign follows a multi-actor dynamic in which the State retains a monopoly. During the communication process, these multiple actors mobilise "cognitive resources" that enable them to produce public policy *narratives* ([125]) aimed at the public and to have an impact on them. In concrete terms, from 2002 onwards, both collective actors (A) and individual actors (B) stepped up communication for behavioural change in the geographical area of Central Cameroon. In this way, Elie is participating in the process of tightening up the modern fight against the malaria

---

[121] See <u>Plan Strategique National de Lutte contre le Paludisme au Cameroun 2007-2010</u>. Op. cit... p. 91

[122] Ibid. p. 92.

[123] Read GERSTLE (J) and PIAR (C), <u>La communication politique</u>. 3ᵉ edition,Paris,Armand Colin, 2016, p. 97.

[124] Ibid. p. 36.

[125] Read ROE (E.M), <u>Narrative Policy Analysis</u>. Durham, Duke University Press, 1994, pp. 36-37.

problem.

A- Greater emphasis on behaviour change communication by collective players

Unlike organisational sociology, which looks at the "actor" in terms of his capacity and his strategic choices in a given context that never completely constrains him, Patrick HASSENTEUFEL considers that the term "actor" "also makes it possible to encompass individuals and groups, since the actor can be either individual or collective"[126] . In other words, while the sociology of organisations focuses on actors in the individual sense, Patrick HASSENTEUFEL teaches us that actors can be both individual and collective. Under the term actor, "we understand an individual (a minister, a member of parliament, a specialist journalist, etc.) or several individuals (forming, for example, an office or a section of an administration) as well as a legal entity (a private company, an association, a trade union, etc.) or a social group (farmers, drug addicts, the homeless, etc.)"[127] . The actor can be both individual and collective. For HASSENTEUFEL, "collective actors" can be administrative entities, government bodies, political organisations, interest groups, communities of experts, etc.[128] . They correspond to "groupings"[129] in the Weberian sense of the term.

The sequence of actions in the fight against malaria, devoted to emphasising communication to change household behaviour, is the work of two types of collective actor: the State and NGOs. These actors are mobilised for a vast communication operation aimed at the general public, the objective being to use discourse to persuade a given listener or audience to take a certain action[130] . To achieve this, they mobilise incentive and communication tools, as well as public policy narratives. In this way, they give new impetus to the traditional fight against the malaria problem by making it tougher. In what follows, we will analyse the mechanisms by which the state (1) and NGOs (2) have stepped up communication to change behaviour.

1. Emphasis on communication to change behaviour by the State

Max WEBER[131] sees the State as a *"political enterprise of an institutional nature"* when and as long as its administrative leadership successfully claims a monopoly on legitimate physical coercion in the application of regulations. What characterises the State is that the "enterprise"[132] that it carries out in a geographical area is accompanied by "physical coercion". In this sense, the State is also a

---

[126] Read HASSENTEUFEL (P), Sociologie politique : l'action publique. Op. cit... p. 172.

[127] Read KNOEPFEL (P), LARRUE (c), VARONE (D) and HILL (M), Public policy analysis, Op. cit... p. 39.

[128] Ibid.

[129] Read WEBER (M), Economie et societe (1922), translated by Pans, Pion, 1991, p. 88.

[130] Read GERSTLE (J) and PIAR (C), La communication politique. Op. cit... p. 108.

[131] See WEBER (M), Ibidem, p. 97.

[132] The concept of "enterprise" comes from Max Weber and refers to a continuous activity carried out by a group organised as an enterprise. This concept obviously also includes the carrying out of political and economic affairs, of affairs specific to associations, insofar as the characteristic of the continuity of an activity with a purpose suits them. See WEBER (M), Economie et societel. Idem. p. 94.

"political grouping"[133] whose social activity is always rationally determined with a view to[134] . It is thus "the universal legatee of unsolved social problems"[135] . In other words, the State only exists in any society on the basis of its actions, which are all aimed at resolving social problems by enacting and implementing "administrative regulations"[136] . In his work, Pierre BIRNBAUM acknowledges the singularity of the State in relation to other actors when he writes: "the State is an actor capable of exerting a specific influence within society, guiding and structuring it"[137] . Thus, the State appears as a "rational institution" or a continuous "enterprise"[138] .

As the "universal legatee of unresolved social problems"[139] , the State has set up a vast communication campaign to change the behaviour of households in the Centre, as a prelude to the mass distribution of LLINs in 2016. This large-scale household awareness campaign complements the campaigns to distribute LLINs to pregnant women and children under the age of five between 2003 and 2007 (we will come back to this in the next section). In order to ensure that future net distribution campaigns did not end in failure, the government decided to carry out an intensive communication campaign prior to any distribution operation. The aim was not only to encourage people to buy LLINs, but above all to encourage them to use them wisely.

The process of raising people's awareness by the State through communication is made plausible by the mobilisation of incentive and communication instruments as well as knowledge resources. During this awareness-raising phase, the state mobilizes several communication channels with a view to having a significant impact on people in the Centre region about the merits of using a mosquito net. In practical terms, the government's emphasis on behaviour change communication was felt in both Yaounde and rural districts.

In the city of Yaounde, a vast campaign to raise awareness among households in the six health districts will begin in October 2015. This Communication Campaign for

Changing Behaviour was designed to sell the merits of the LLINs to be distributed in 2016. It was therefore up to the government to mobilise the population by encouraging them to take part in the net distribution process through a large-scale communication campaign aimed at explaining the operation to them. Hence the use of incentive and communication tools. In this way, the State was able to mobilise

---

[133] This is because its existence and the validity of its regulations are guaranteed by the continued existence within a geographical territory that can be determined by the application and threat of physical constraint on the part of the administrative management. See WEBER (M), Ibid. pp. 96-97.

[134] Ibid. P. 68.

[135] See BIRNBAUM (P) and BADIE (B), Sociologie de 1 Etat. Pans, Grasset, 1979, p. 30.

[136] These are all rules which apply to the behaviour of the administrative management as well as, to use an expression which has become commonplace, to the behaviour of the members "towards the group". See WEBER (M), Idem. p. 93.

[137]Read BIRNBAUM (P) and BADIE (B), Sociologie de 1 Etat. Op. cit... p. 61.

[138]Read WEBER (M), Op. cit... p. 99.

[139] See BIRBAUM (P) and BADIE (P), Idem, p. 30.

several communication channels to raise awareness and mobilise the population before starting the process of distributing LLINs. In the city of Yaounde, for example, a study showed that out of 1006 households, 59.8% had been informed about the 2016 distribution of LLINs[142] . This behaviour change communication campaign was carried out through various communication channels and the results obtained are shown in table n° 3 below.

Table 3

Percentage of households informed of the
2016 distribution of LLINs
in Yaounde, by communication channel

Communication channels	Percentage of households informed about the distribution of LLINs in 2015/2016
Radio	8,8%
Television	30,3%
Banner	17,8%
Hospital/health centre	6,6%
Community health worker	13,9%
Traditional/administrative authority	4,0%
Church/mosque	9,5%
Home visit by the enumeration/awareness team	16,4%
Family/friend	18,5%
Voisin	19,5%
Journal	4,6%
Telephone	16,6%

[142] Read the document entitled <u>Cameroon: enquete post campagne sur l'utilisation des moustiquares impregnees d'insecticide a longue duree d'action 2016/2017</u>. Rapport final, MINSANTE, PNLP, INS, Decembre 2017, p. 71.

Other	0,3%

Source : final report, MINSANTE, PNLP, INS, <u>Cameroun : enquete post campagne sur l'utilisation des moustiquares impregnees d'insecticide a longue durée d'action 2016/2017,</u> decembre 2017, p. 7

We can see that the State has mobilised both communication tools (television, radio, banners, newspapers, telephone, etc.) and incentives (community health workers, traditional and administrative authorities, families and friends, neighbours, health facilities, etc.) to successfully carry out the campaign to raise awareness among the population about the mass distribution of LLINs in 2016. No fewer than thirteen communication channels were mobilised, with the aim of having a considerable impact on local people through "direct persuasion"[140] .

---

[140] Persuasion is direct when the message controlled by the actor or disseminated by the media modifies an individual's attitude towards a reform, a political image or any other publicised, politicised and polarised object by

47

Television, neighbours and banners were the communication channels through which most of the population was informed of the forthcoming LLIN distribution campaign. According to official figures, out of 1006 households living in Yaounde, 59.8% were informed about the distribution of LLINs before the operation began[141] . This meant that a large proportion of Yaounde's population was already prepared to receive an LLIN. The implementation of this social behaviour change communication campaign will be presented and analysed in the remainder of this report.

In the twenty-four health districts on the outskirts of Yaounde, the State also carried out an intense communication campaign about the forthcoming LLIN distribution campaign, similar to the one in Yaounde. However, in these social areas, it used incentives rather than communication tools, with much greater results, as 78.2% of a sample of 491 households had been informed about the 2016 LLIN distribution campaign[142] since October 2015, well before the process was launched. The communication campaign in the outlying areas was also carried out through various communication channels, and the results obtained can be seen in table° 4 below.

Table 4

**Percentage of households informed of the
2016 distribution of LLINs
in the twenty-four peripheral districts, by
communication channel[146]**

Communication channels	Percentage of households informed about the distribution of LLINs in 2015/2016
Radio	5,5%
Television	6,9%
Banner	1,5%
Hospital/health centre	22,6%
Community health worker	41,9%
Auto rite traditional/administrative	30,5%
Church/mosque	18,6%
Home visit by the enumeration/awareness team	40,5%
Family/friend	44,1%
Voisin	40,5%
Journal	6,8%
Telephone	5,7%

adding information to the stock of considerations available in the memory. See GERSTLE (J) and PIAR (C), La communication politique. Idem, p. 108.

[141] Read the document entitled Cameroon: enquete post campagne sur l'utilisation des moustiquaires impregnees d'insecticide a longue duree d'action 2016/2017, Rapport final, MINSANTE, PNLP, INS, Decembre 2017, p. 71.

[142] Ibid.

| **Other** | 0,4% |

Source: final report, MINSANTE, PNLP, INS, <u>Cameroon: post-campaign survey on the use of long-lasting insecticidal nets 2016/2017,</u> December 2017, p. 7

The State's preferred channels for raising awareness among households in rural areas were "neighbours", "family", "home visits by the awareness team", "community health workers" and "traditional authorities". Traditional communication channels such as "television", "banners", "telephone" and "radio" did not have enough impact on households in outlying areas. In these social areas, incentives have become the norm for raising people's awareness of the upcoming LLIN distribution campaign. This is due to the fact that the districts on the periphery are enclosed (conventional means of communication y are difficult to mobilise), small in size and characterised by strong primary solidarity and communitarianism. The state has therefore made the community the focus of its communication campaign in rural areas. Consequently, "family/friends", "neighbours", "traditional authorities" and "community health workers" have been reliable channels for transmitting messages to rural populations because of their sociological roots in rural areas. From this perspective, behaviour change communication was based on the "*two-step* flow of communication" model[143] . This model proposes a dynamic of the influence of the media through interpersonal relationships. The influence of the media in opinion formation is exerted through opinion leaders[144] . In the rural areas of the Centre region, behaviour change communication could only be plausible if it materialised through these opinion *leaders,* who in this case were the traditional authorities, neighbours, community health workers and families. Since they share the same social determinants as the local population, they are in a better position to convey messages about malaria prevention and facilitate their acceptance than the traditional media, which are mostly based in Yaounde. According to official figures, 78.2% of a sample of 491 households had been informed about the distribution of 2016[145] LLINs since October 2015, well before the process was launched. The percentage of households informed was much higher in rural areas than in Yaounde. In rural areas, the government has taken on board the principle that "for communication to be effective, the message must not only conform to the values of the groups to which the recipients belong, but must also be supported by an opinion leader"[146] . This is why the results obtained here in terms of awareness-raising were very satisfactory, as more than half the households already knew what to expect and were prepared to receive and use the nets.

---

[143] Read KATZ (E) and LAZARDSFELD (P), <u>Influence personnels.</u> Pans, Armand Colin, CEFAI Daniel, 2008.

[144] Read GINGRAS (A-M), "Les theories en communication politique", in GINGRAS (A-M), (dir.) <u>La communication politique. Etat des savoirs. enjeux et perspectives.</u> Presse de I'Universite du Quebec, 2003, p. 17.

[145] Read the document entitled <u>Cameroon: enquete post campagne sur l'utilisation des moustiquares impregnees d'insecticide a longue duree d'action 2016/2017.</u> Rapport final, MINSANTE, PNLP, INS, Decembre 2017, p. 71.

[146] Read GINGRAS (A-M), "Les theories en communication politique", in GINGRAS (A-M), (dir.) <u>La communication politique. Etat des savoirs. enjeux et perspectives.</u> Idem, p. 18

Ultimately, the behaviour change communication campaigns organised by the State between October 2015 and January 2016, both in the centre of Yaounde and in rural areas, were part of a drive to step up the modern fight against malaria. They were a prelude to the vast campaign to distribute LLINs free of charge on a mass scale. To achieve this, households had to be "prepared" to receive this vast public action programme. To do this, the government used incentives and communication tools to take considerable action in the social arena. Several communication channels were used to inform households of the importance of using a mosquito net. Alongside the State, NGOs have also become involved in the process of emphasizing communication to change household behaviour in the Centre-Cameroon region.

2. Emphasis on behaviour change communication by local NGOs

The Centre region is not the victim of what Clement SORIAT calls an "associative pathology"[147] because from 2002, with the adoption of the *"Roll Back Malaria"* programme, a multitude of local NGOs emerged in this geographical area[148] . This is simply the logical consequence of the Cameroon government's desire to stage its "democratic transition" and open up the political game. Today, local NGOs play a relatively institutionalised role and appear to be partners in public action, under the control of state actors and international donors acting through the state.

"delegation"[149] . Their rise is merely the consequence of "the structuring of multi-actor public action"[150] .

According to Edrich Nathanael TSOTSA, the term NGO seems socially and even politically more correct and more prestigious than the term association, both in the eyes of political partners and external donors[151] . He goes on to explain the difference between an NGO and an association in the following terms: "the term NGO is used to designate all the large-scale organisations that enrich the logic of civil society, whereas the term association refers to small-scale structures of lesser visibility"[152] . To make this distinction clearer, the author adds that "an association is an elementary structure which, in Burkina Faso and in French-speaking Cameroon, is referred to as a club, amical, mutuelle or mouvement. On the other hand, an NGO y appears to be a socially stable organisational structure, with large human resources and financially viable and autonomous"[153] . It would appear that the difference between an NGO and an association lies in the fact that the former is much more active in the social field than the latter. In the Centre region, for

---

[147] Read SORIAT (C), "L'implication des acteurs associatifs beninois dans 1 action publique de lutte contre le sida : entre domestication et prise de pouvoir", in <u>"Penser l'action publique en contexte africain"</u>. Congres AFSP Aix, 2015, p. 2.

[148] See the list of all the NGOs working in the field of malaria in the Centre region in appendix 2 of this report.

[149] Read SORIAT (C), Op. cit... p. 2

[150] Ibid., p. 4.

[151] TSOTSA (E.N) <u>L'action publique de lutte contre le VIH/Sida. Actors, controversies and dynamics. Analyse comparee a partir des exemples sud-africains. burkinabe et camerounais</u>. These de doctoral en science politique, Institut d'Etudes politiques de Bordeaux, 2009, p. 323.

[152] Ibid. p. 323.

[153] Ibid. pp. 323-324.

example, NGOs are private actors who act as a "safety net" for many families affected by malaria and left to fend for themselves.

In the Centre-Cameroon region, the process of emphasising behaviour change communication also involves local NGOs. These are also collective actors[154] who carry out "therapeutic activism" by intervening in the field of malaria through large-scale household awareness campaigns. They act as a complement to government action and help to step up the fight against malaria in the Centre region. Elies mobilises human and knowledge resources, stories and public policy statements to take significant action in the social field. The focus of the NGOs' communicative action is on the need to use mosquito nets in order to prevent malaria. Like the government, local NGOs see awareness-raising as a prerequisite for free mass distribution of ITNs/MILDAs. The aim is to appeal to households and prepare them to receive and use ITNs/MILDAs in order to avoid misuse of this instrument in the fight against malaria. In the Centre region, three NGOs can be singled out as representatives of the implementation of behaviour change communication among the panoply of NGOs that exist in this geographical area[155]. Three NGOs stand out for their activism in the field of social communication to raise awareness of the use of mosquito nets: the Association Camerounaise pour le Marketing Social (ACMS), Malaria no more (MNM) and Impact Sante Afrique (ISA).

While ACMS' traditional activities focused on the marketing and distribution of "super mosquito nets" and "Bloc" impregnation kits to the general public, it should be noted that between 2002 and 2005, ACMS was much more involved in large-scale awareness-raising through communication. During this period, ACMS launched its major household awareness campaign, the "Behaviour Change Communication Campaign" (BCC). It is the translation on the ground of the policy known as   "CCC is the cognitive resource through which   CASI manages to have a significant impact on people's behaviour by using the various communication media to which they are constantly exposed. Thus, during the 2002-2005 period, the NGO conveyed malaria-related messages through various communication media, namely television, radio, billboards, posters, leaflets, health centres and ACMS/Blitz agents. A field survey carried out in the city of Yaounde by IRESCO (Institut de Recherche et des Etudes de Behaviour) showed that the population in this sub-system had been exposed to various messages from ACMS between 2002 and 2005 as part of the CCC[156]. The messages conveyed to the public were as follows:

- "malaria is transmitted by mosquito bites at night. Protect yourself by using an insecticide-impregnated mosquito net"[157] ;

- "pregnant women and children under 5 are the people most at risk of malaria.

---

[154] In the sense that they are non-governmental organisations and therefore legal entities.

[155] See the list of NGOs working in the field of malaria in the Centre region in appendix 3 of this report.

[156] See the document entitled Etude d'évaluation du projet de distribution des moustiquares impregnees a l'insecticide par l'association camerounaise pour marketing social. Document prepared by l'Institut de

Protect them by using an insecticide-impregnated mosquito net"[158] ;

- "super mosquito net, a gift for life"[163] ;

- "reinforce the effectiveness of your mosquito net every 6 months by soaking it in a mixture of water and Bloc tablet "[164] ;

- "Block, the strength of your mosquito net"[165] .

Billboards have been erected at major crossroads in Yaounde to convey these messages, such as the "Bata Longkak" crossroads[166] . Leaflets were distributed in the streets by a team of young ACMS/Blitz agents; posters were put up on the walls of health centres; local radio stations were mobilised to broadcast three advertising spots produced in *"Mongo Ewondo"* to ensure that the message was better perceived and received by people from Central Cameroon.

Messages with "strong" and "precise" content were conveyed by ACMS to the population of Yaounde and the surrounding area using various communication media. To take the first message as an example *("Malaria is transmitted by mosquito bites at night. Protect yourself by using an insecticide-impregnated mosquito net"),* a study carried out in February 2006 showed that, out of a sample of 395 people living in Yaounde, 47.3% received the message from television, 38.2% from the radio, 15.9% from billboards, 22.0% from posters, 3.8% from leaflets, 29.6% from health centres and 3.3% from ACMS/Blitz agents. The same study showed that 32.4% of people had not heard or seen the advertisement[159] . These figures alone demonstrate that ACMS has done a "titanic" job in the field of communication by popularising methods of combating malaria using mosquito nets. Aware that the "message is the medium"[160] , television, radio, health centres and posters were the preferred media through which ACMS's Behaviour Change Communication Campaign was made plausible between 2002 and 2005. Households were thus prepared to receive and use a mosquito net.

Malaria no more (MNM)[161] and Impact Sante Afrique (ISA) are two NGOs that are also involved in the process of stepping up communication to change household behaviour in the Centre-Cameroon region. Elies share the same Cameroon executive management team, based in Yaounde. Since 2015, Elies has been working in the field of malaria by organising household awareness campaigns as part of a global policy known as "Communication for Behaviour Change". Elies

Recherche et des Etudes de Comportement (IRESCO), February 2006. Also available online at www.yumpu.com.

[161] Ibid. p. 22

[162] Ibid. p. 22.

[163] Ibid. p. 23

[164] Ibid. p. 23.

[165] Ibid. p. 23.

[166] Poster that still exists today.

[159] Figures obtained from document entitled Etude d'évaluation du projet de distribution des moustiquaires impregnees a l'insecticide by l'association camerounaise pour le marketing social. Ibid.

[160] See Me LUHAN (M), Pour comprendre les medias. 1964, Marshall Me LUHAN, transl. fr. 1968, reed. Seuil, "Points essais" collection, 1977.

[161] This NGO was founded in Seattle in the United States in December 2006 by Peter CHERNIN and Raymond CHAMBERS. In Cameroon, this NGO is based in Yaounde and is a leading partner of MINSANTE in the fight against malaria.

therefore uses knowledge resources, stories and public policy statements to act effectively in the social field.

With the aim of launching a major public awareness campaign, the NGO MNM organised a workshop in Yaounde in 2015 to build the capacity of parliamentarians in the prevention and management of malaria in Cameroon. The aim of the workshop was to set up a public awareness campaign by involving the MPs trained by MNM in the process. For the NGO, adequate and effective awareness-raising among households could only plausibly be achieved by involving the "opinion *leaders*"[162] referred to by Paul Lazarsfeld, in the process. As MPs are closer to the people (because they are elected by them within the department), they may be in a better position to raise their awareness through face-to-face communication. It was at the end of this training workshop that parliamentarians from the Centre region pledged to be "ambassadors in the battle against malaria", led by Marie-Rose NGUINI EFFA, MP for Mefou-et-Akono. We will come back to this in the next paragraph. The least we can say at the moment is that MNM has done a titanic job in the field of communication through the Executive Director of this NGO. No fewer than fifteen radio programmes were produced between November 2015 and January 2016 by Olivia Ngou, half of which were broadcast on Crtv-Radio-Centre[171] . The message conveyed in these broadcasts was about practical ways of using impregnated mosquito nets.

Between 2018-2019, 1 NGO ISA is also involved in raising public awareness through strong "therapeutic activism". Elie called on communities to get actively involved in the fight against malaria. The members of this organisation carried out concrete persuasion work aimed at the population, its sole and main target *group*. This is the case of Mrs NGO BATJE Henriette who, since 2018, has been making speeches to students in the Nyong-et-Kelle department, who are members of the "Nyong-et-Kelle Students' Association" and live in Yaounde. One of ISA's fundamental missions is to "engage influential members of civil society and communities so that they can become actively involved in the fight against malaria"[172] . Students from the Universities of Yaounde 1 and 2, as well as from a number of private higher education institutes, were fully involved in awareness campaigns in the villages of Matomb and Boumnyebel in August 2019. The message conveyed to these communities focused on basic hygiene measures and practical methods for using LLINs. Mrs Nafissa Tame did the same for Muslim women's associations in the Centre region, and in Yaounde in particular. This awareness-raising work, which leads to alliances, corresponds to what Philippe

---

[162] Read BALLE (F), <u>Medias et societe</u>. Paris, Montchrestien, 9th ed. 1999.
[171] Interview with Olivia Ngou, Cameroon Executive Director of the NGO malaria no more on 30 November 2020.
[172] Ibid.

ZITTOUN calls "public policy statements"[163] which are discursive practices based on persuasion and argumentation by actors to get others to agree with the problem to be solved. By producing statements, ISA agents were able to build the 'alliances', 'associations' and even 'coalitions' needed with a large section of the population to combat malaria, which requires a high level of participation from the population as a whole, in accordance with the principle of 'multisectoriality'. By "making statements", the NGO ISA has managed to exert "power" over households, who have ended up getting involved in the fight against the pandemic by using LLINs.

In addition, as part of the use of its knowledge resources, the NGO ISA mobilised the Yaounde media to produce stories aimed at the local population. Olivia Ngou, the NGO's Executive Director, was constantly on the air on local radio stations Magic FM, Royal FM and Crtv station regionale du Centre with the following message: "*sleep and have your children sleep under an impregnated mosquito net to avoid the risk of malaria*"[164] . The aim of this media campaign was to encourage people to take a number of steps to prevent the risk of malaria.

In short, the collective actors, i.e. the State and the NGOs, are effectively involved in the process of raising awareness among nationals. On this occasion, they mobilised several instruments and resources to better implement the Behaviour Change Communication Campaign in the social field. They are being followed by individual players.

B- Emphasis on behaviour change communication by individual players

The emphasis on communication to change people's behaviour is not the prerogative of collective actors alone, as individual actors are also involved in the process. In the Centre region, a large number of actors are involved in the field of malaria in addition to the communication activities of the state and NGOs. For the most part, these are individual actors who acquire the status of actor simply by belonging to our field of study. These are individuals whose actions in favour of malaria are palpable in the social arena. In their own way, they contribute to the intensification of the modern fight against malaria.

Unlike collective actors, individuals are essentially rational and strategic, i.e. they appear in a public policy process once their interests come into play and they have the resources to defend them[165] . His approach will always make sense to him in terms of his objectives and in terms of the opportunities he sees and chooses to seize[166] . The actions of individual players in the field of malaria in Central Cameroon are not devoid of interest. They have interests to satisfy, and to do so they mobilise resources of various kinds that enable them to have an impact on the malaria disease. Analysis of their actions in the field of malaria must therefore

---

[163] See ZITTOUN (P), "La fabrique pragmatique des politiques publiques", Op. cit... p. 79.

[164] Interview with Olivia Ngou, Cameroon Executive Director of the NGO malaria no more on 30 November 2020.

[165] Read KNOEPFEL (P), LARRUE (C), VARONE (F) and HILL (M), Public Policy Analysis, British Library, 2007, p. 40.

[166] See GROSSMAN (E), "Actor", Op. cit... p. 26.

imperatively take account of the resources they mobilise and the interests they defend. Individual actors in the field of malaria in the geographical area of the Centre include political and administrative elites and intermittent actors. They became involved in the process of stepping up behaviour change communication from 2002 onwards. We will therefore analyse the way in which grassroots political and administrative elites (**1**) and so-called intermittent actors (**2**) put behaviour change communication into practice in the social arena.

1. Emphasis on behaviour change communication by grassroots political and administrative elites

Anglo-American research in political science offers a variety of definitions of the concept of elites. Anthony GIDDENS considers elites to be "individuals who occupy formally defined positions of authority at the head of a social or functional organisation"[167] . For Thomas BOTTOMORE, "elites designate those functional groups which, for whatever reason, occupy a high social rank"[168] . Robert PUTNAM considers elites to be "individuals who have the ability to influence political decisions"[169] . As for Ezra SULEIMAN, "all people who occupy positions of authority belong to the elite"[170] . Defined in this way, the concept of the elite does not designate a precise category of individuals, since depending on the context and circumstances, one or other individual may take advantage of the attributes of the elite. An elite can therefore be political, economic, judicial, civil servant, military, administrative, etc. But for the purposes of this paragraph, we will only refer to the category of elites known as political-administrative.

By the expression "grassroots political and administrative elites", we mean all the political and administrative players who are close to the people and who are better able than anyone else to have a direct impact on their daily lives. The expression is borrowed from Michael LIPSKY, who in one of his seminal works spoke of *"street-level bureaucrats"*[171] to designate the grassroots or field agents of the administration[172] . He was referring to civil servants *in the strict sense of the term*, who have a direct influence on the citizens for whom a public policy is intended. Unlike the concept of *'street-level bureaucrats'*, the expression 'grassroots politico-administrative elites' refers not only to the category of actors known as 'civil servants', but also to political actors. It can thus be used to designate any actor who maintains 'face-to-face' contact with the public and whose actions have direct effects on the existence of the individuals affected by public policy. In this sense, the concept of 'grassroots politico-administrative elites' makes it possible to render

---

[167] Quoted in GENIEYS (W), <u>Sociologie politique des elites</u>. Paris, Armand Colin, 2011.
[168] Ibid.
[180] Ibid.
[179] Ibid.

[181] Read Lipsky (M), <u>Street-Level Bureaucracy: Dilemmas of the Individual in Public Services</u>. New York, Russel Sage Foundation, 1980.
[172] See HASSENTEUFEL (P), <u>Sociologie politique : l'action publique</u>. Op. cit... p. 159.

plausible what Ranate MAYNTZ calls 'an implementation structure'[173] - in other words, the reality of the phenomena, the concrete face as it emerges from implementation on the ground. From this point of view, a set of relationships is established between players. Problems are experienced and criteria are used in a certain way[174] . In other words, in the process of implementing a public policy, the 'grassroots political-administrative agents' precede the field through the mechanism of 'adaptation to circumstances'[175] which enables them to adjust a public decision according to the local situations they face.

Thus, to analyse the intensification of the fight against malaria on the basis of the actions of "grassroots political and administrative elites" is still to grasp public action on malaria "from below", highlighting the issue of proximity in politics. Christian LE BART and Remy LEFEBRE[176] describe the heuristic interest of the notion of "proximity" in the analysis of public policy in the following terms: "Through proximity, it is a question of restoring political confidence (bringing elected representatives closer to citizens), re-establishing social links (bringing institutions closer to users and users closer to each other), and rebuilding public efficiency (sticking to social demand, producing appropriate, adjusted responses)". In other words, the notion of proximity fits in with that of territorialisation, making it possible to analyse local public policies.

In the Centre region, local public policies are identified through the actions of grassroots political and administrative agents. It is these actors who adjust malaria control policy in the field according to the specific situations they face. They make it possible to territorialise public action against malaria in this geographical area. They are elites who mobilise political, social and temporal resources, and whose actions are strategic because they have interests to defend. The Behaviour Change Communication Campaign implemented in the Centre region since 2002 involves individual actors who desectorialise and de-institutionalise the fight against malaria: administrative and political elites.

The involvement of administrative elites in the field of malaria in the geographical area of Centre-Cameroon began in 2002, in accordance with the principle of "multisectoriality". Questioning the actions of these so-called administrative elites also means taking into account the role of grassroots or 'field' agents of the administration in the process of implementing public action. Michael LIPSKY refers to them as *'Street-Level Bureaucrats'*[177] , civil servants who participate in the construction of public action in a sub-system. As a result, like elected representatives, they are players in the territorialisation of public policy, by virtue

---

[173] Read MENY (Y) and THOENIG (J-C), <u>Politiques publiques.</u> Op. cit... p. 254.
[174] Ibid. p. 254.
[175] Ibid. p. 274.
[176] Read LE BART (C) and LEFEBRE
Presses Universitaires de Rennes, Rennes, p. 29 (R), <u>La proximite en politique. Usages, rhetoriques. pratiques.</u>

[177] Read LIPSKY (M), <u>Street-Level Bureaucracy: Dilemmas of the Individual in Public Services</u>. Op. cit... pp. 3-23.

of their proximity to citizens. From the work of Max Weber to the present day, the definition of the concept of civil servant has undergone a heuristic evolution. The true civil servant," said Max Weber, "must not engage in politics, precisely by virtue of his vocation: he must administer, above all in a non-partisan manner"[178] . In the Weberian sense of the term, the civil servant is characterised by non-commitment, neutrality and dehumanisation. Yet recent work in administrative science shows that civil servants are highly politicised, hence the concept of the "functional politicisation" of civil servants[179] . In this sense, the loyalty of civil servants to their superiors encourages them to be committed and influential. Jean-Michel Eymeri writes that: "the institution assigns the senior civil servant the task of convincing himself of things and convincing the politician of them, of committing himself and committing the politician to committing himself: he must exercise a ministry of influence"[180] . Civil servants therefore cease to be considered as "functions in action"[181] because they have a bias and interests to defend, and consequently their loyalty becomes a "duty of commitment"[182] .

The brief presentation of a revolution in the concept of the civil servant was undeniably necessary in order to understand the position of senior civil servants in the process of implementing the 'Behaviour Change Communication' policy in the Centre region. Far from being perfectly neutral actors in the Weberian sense of the term, senior civil servants have interests to defend and protect. They act on behalf of their superiors and wish to benefit from their trust in order to be promoted to more important positions. Their involvement in the process of stepping up the fight against malaria by raising awareness among households becomes a pretext for them to express their loyalty to their superiors. This is the whole point of their involvement in the field of malaria from 2002 onwards. In this sense, their actions are essentially strategic and self-interested.

In the geographical area of Centre-Cameroon, the basic administrative elites who act in the field of malaria by emphasising communication for behavioural change are the civil servants of the territorial administration, i.e. the administrative and traditional authorities. In other words, civil administrators and their auxiliaries, traditional chiefs. They intervene in the field of malaria through symbolic policies by producing speeches and stories aimed at the population. It is precisely at this level that the symbolic dimension of public action is taken into account. Patrick HASSENTEUFEL argues that: a l action publique comprend une importante dimension symbolique a travers les discours et les actes de communication qui l'accompagnent. They do not necessarily refer to concrete measures, but they can

---

[178] WEBER (M), <u>Le savant et le politique</u>. Pion, 1959, p. 108.
[179] To design this consideration of the political dimension of their activities by senior civil servants in the administration.
[180] Read EYMERI (J-M), "Frontiere ou marches? De la contribution de la haute administration a la production du politique", in LAGROYE (J) (ed.), <u>La politisation</u>. Paris, Belin, 2003, p. 70.
[181] Read BIRNBAUM (P) and BADIE (B), <u>Sociologie de 1 Etat.</u> Op. cit... p. 186.
[182] EYMERI (J-M), Idem. p. 71.

modify public perceptions, expectations and even behaviour"[183] . From this point of view, the signifier (the meaning given to the action) outweighs the signified (the action as such)[184] . In practice, in the Centre region, civil administrators and traditional chiefs are involved in the process of stepping up the fight against malaria through communication about the disease. Since 2003, Emile Onanbele Zibi, President of the indigenous patriarchs of Mfoundi, has been involved in the fight against malaria through communication. His position as a patriarch with "traditional power" was a social resource for him and made it easier for the population to accept and internalise his message. During the distribution of the Mil to pregnant women in the Centre region in 2003, 2004 and 2005, this administrative elite (auxiliaries to the administration) took centre stage in raising awareness among the target audience, represented here by the pregnant women of the Febe-village neighbourhood, of the need to equip themselves with a mosquito net. A woman called Marylise, a saleswoman at the Mfoundi market and resident in the Febe-village neighbourhood, testified in an interview:

"The patriarch Onambele, peace be upon him, came to our house once. He told me about the mosquito net, saying that it was what was going to save us from malaria and that they were going to come and distribute it to us *ða* soon. He also said not to close the door during the day and to get

ii 195

the people who are going to come with the mosquito nets".

These are the words of a woman whose name appeared in the directory of community health workers in Mfoundi and who was to receive free MU during the 2003 campaign. It would appear that the patriarch ONAMBELE ZIBI himself went out into the field to raise awareness among pregnant women and encourage them to take and use MU. It's worth noting that he targeted a few households near his home in the Febe-village neighbourhood to do this, and asked his peers to do the same in their neighbourhoods. This communication campaign continued until the major distribution of LLINs began in 2016. We can therefore say that the success of the LLIN distribution campaigns in Yaounde (more so than in the surrounding areas) was partly the work of this administrative elite. We will come back to this in the next section.

The civil administrators of the Mfoundi department have also played an important role in the process of stepping up communication to change the behaviour of the people of Yaounde since 2004. Prior to the campaign to distribute Mils to pregnant women in the Centre region, Prefect Pascal Mani was personally involved in the awareness campaign, visiting a number of homes in the Longkak and Tsinga neighbourhoods to encourage pregnant women to take[185] the Mils distributed to them and make good use of them. The awareness-raising process gathered pace from 2011, with the implementation of the *'K.O Malaria'* programme, which is essentially based on the large-scale distribution of Mil to the population. Prefect

---

[183] See HASSENTEUFEL (P), <u>Sociologie politique : Faction publique</u>. Op. cit... p. 61.
[184] Ibid.
[195] Interview with Marylise, trader at the Mfoundi market, 13 February 2021.

Jean-Claude TSILA was personally involved in the campaign to raise awareness among the population prior to the mass distribution of ITNs/MILDAs in 2016. A study carried out in the Centre region during the 2015/2016 LLIN campaign showed that out of 78.2% of households reached, 30.5% had been reached by the administrative and traditional authorities[186] and in the city of Yaounde, 4% of households had been reached by them[187]. Furthermore, in the run-up to the World Malaria Day, a special malaria awareness and advocacy day was held in Yaounde on Wednesday 17 April 2019. Organised by the French Embassy and the NGO Impact Sante Afrique, the Yaounde administrative authorities took part and took centre stage. The aim of this special event was to raise the profile of the LLIN among the local population. In the Tsinga district in the centre of Yaounde, a demonstration of the use of LLINs was organised by the Prefect in front of an audience of health professionals: this was the materialisation of the "symbolic dimension of the communication process"[188]. Subsequently, the Minister of Public Health, the governor of the Centre region, the Prefect of Mfoundi and the sub-prefect of the Yaounde 2 district distributed the LLINs to the local residents. In this way, the "chiefs of the land" were the architects of symbolic policies aimed at the local population. Not a single sub-prefect was not involved in the campaign to raise awareness among households about the use of mosquito nets between 2015-2016[189].

In addition, as far as the political elites are concerned, their involvement in the field of malaria began in 2002, in accordance with the principle of "multisectoriality". Like the State and NGOs, they are known for organising communication campaigns to change behaviour by producing stories for the general public. In 2015, the majority of the political elites in the Centre region made a solemn commitment to be "ambassadors for the fight against malaria"[190]. The most emblematic and charismatic political figure in this exercise was Marie-Rose NGUINI EFFA, MP for Mefou-et-Akono. Her position as president of the "network of parliamentarians, populations and development" and as a member of the "association of dynamic women of Cameroon" gives her much greater visibility in geographical areas other than her electoral constituency. As a result, she has constantly produced reports on the politics of malaria for the people of Yaounde in particular and the Centre region in general. The reports in question focused on ways of preventing malaria in pregnant women and children through the use of mosquito nets. Her position as an elected representative of the people helped to make her speech more operational, a real political resource for her. Elie thus played

---

[186] See the document entitled <u>Enquete post campagne sur 1 utilisation des MILDA 2016/2017. Rapport final</u>. Op. cit... p. 70.

[187] Ibid. pp. 58-59.

[188] Read GERSTLE (G) and PIAR (C), <u>La communication politique</u>. Op. cit... pp. 25-29.

[189] Interview with Mr Serge Herve BIWELE SAL, sub-prefect of the Yaounde 3 district[eme] on 13 April 2021.

[190] During a training workshop organised by the NGO Malaria no more in 2015 to build the capacity of parliamentarians in the Centre region to prevent malaria.

a major role in encouraging the people of Yaounde to take up the LLINs distributed by the State in 2016. We will come back to this in the next section. It is in this sense that stories are to be considered as an important cognitive resource used by an actor to better convey a given public action programme[191] .

What's more, MP Marie-Rose NGUINI EFFA ran a major media advertising campaign on the subject of malaria. The advert in question was entitled *"K.O paludisme, tous unis contre le paludisme" (malaria, all united against malaria)* and was broadcast in all the major media outlets broadcasting in the city of Yaounde since 2014. The message conveyed by this advert was about "protecting l'enfant against malaria". This political elite insists on the use of MU by parents to protect their children. The advert, which featured the MP in person, was widely broadcast on Crtv-tele, Canal 2 international, Crtv's national station, Crtv's regional station for the Centre, Royal FM and even on social networking sites such as *You Tube.* In this way, the MP's message had a significant impact on the behaviour of the population (target group) with regard to the use of mosquito nets (we will come back to this in more detail in chapter 4 of this report). The fact that MINSANTE chose a political elite to convey this advertising message is not insignificant. It was a question of using the political and social resources available to Mrs NGUINI EFFA to influence the population in a positive direction on the subject of malaria, and in this case on the use of LLINs. The actions of political actors in the field of malaria are not limited to the electoral period. It is therefore not only strategic, but also highly humanitarian.

2. Emphasis on behaviour change communication by intermittent players
According to Patrick HASSENTEUFEL, the intermittent actors of public action, who maintain an excessively close relationship with the media, are the journalists and individuals (experts, intellectuals, politicians, etc.) who express themselves in the media[192] . They convey messages that can have a direct effect on the public. Through their constant interventions in the media, "intermittent actors" modify people's public perceptions of an attitudinal object. In this sense, they are crucial players in the process of stepping up the fight against malaria. In the Centre region, since 2010, intermittent actors have been veritable 'artisans' of the materialisation of behaviour change communication on the social ground through their constant interventions in the national and local media. Their speeches focus mainly on ways of preventing malaria, particularly through the use of LLINs. The intermittent actors involved in the field of malaria are musicians and socio-political actors.
From 2010 onwards, musician-artists entered the media field to play their part in the process of raising awareness among households and thereby stepping up the fight against malaria. Radio and television are the media tools used by these intermittent actors to reach a wide audience. This was the case with the media

---

[191] Read KUBLER (D) and DE MAILLARD (J), <u>Analyser les politiques publiques</u>. Op. cit... p. 166.
[192] Read HASSENTEUFEL (P), <u>Sociologie politique : l'action publique</u>. Op. cit... p. 192.

message broadcast as part of the *"Nightwatch"* programme[193] , in which the voice of the musician Richard BONA could be heard saying: *"It's 9pm, are you and your family sleeping under your mosquito net tonight?* This message was broadcast every evening in French and English on the Crtv national station, the Crtv regional station for the Centre, FM 94 and Magic FM. Its aim was to encourage people to get hold of a mosquito net and use it before going to sleep at night. Artist Richard BONA is urging households to use a mosquito net, while at the same time preparing them psychologically for the major campaign to distribute LLINs, which is just around the corner.

In the same vein, a hymn entitled "K.O malaria"[194] was broadcast over and over again on the local audiovisual media in the Centre-Cameroon region. Every day, households were able to see on television and hear on the radio artists such as Ottou Marcelin, Petit Pays, Lady Ponce, Richard Bona and Sine Turn raise awareness of the disease through the following song:

"We know what to do
We won't let malaria stop us now
Follow your dreams and fight H
Malaria must get out
Listen to this pose, people are talking
You know you forgot mama
To take your second dose of treatment
Pregnant woman, pregnant woman
If your child could only speak
He'll be able to tell you mamaVas a 1 hopital mama
That's all you have to do
Vas a 1'hopital vas a 1'hopital mama
Malaria...
REFRAIN:
We know what to do
Don't let malaria stop you
Pursue your dreams and be yourself
Malaria must go
Your baby has transplant fever at night
Mere Mere inquiete you hope he'll be okay
Don't delay, follow what you know
Go and do the treatment
Don't be the Aye Aye malaria e
Malaria, Wa ngwa ma ndol'a nje
CHORUS
Listen to this bass
You know you forgot
To take your second dose of treatment
Pregnant woman, if your child could talk he would tell you
Go to hospital mum

---

[193] This programme, initiated by the "K.O malaria" campaign, is the result of a vast media operation on both television and radio, the aim of which is to encourage people to sleep under a mosquito net.
[194] Anthem recorded in 2011.

Can you feel it
The music's playing everywhere you go
Malaria treatment no forget your dose
To all the pregnant mamas
I no want me palapa
So go on the clinic take your drugs
Every body listen
We just need prevention
Change the situation bed nets for the nation
Malaria's a killer don't take it lightly
Many people 're dying bed nets hanged nightly
REFRAIN
Your boy has fever
He is sweating through the night
Don't wait more, Hoping he'll be fine
Get a test now threat him don't be slow
Don't delay follow' what you know
Do it so quick, malaria must no go e
REFRAIN
Malaria malaria...
Yo ma dibo wa aye, yo ma dibo wa ye
E ma mbamba na ndinga mwa o a na mwese ya e biye
Ayo ooyoyo yo
Can you feel it
CHORUS
Yeah yeah yeah yo ma dibo wa ma mbamba
Ndutu monamo kolo guie elangi ti monamo
Na mwesse na i biye Malaria malaria na wo, Myola ngo
Follow your dreams and fight it, Malaria ... 205
must get out... " .

This hymn, which is an exhortation to fight malaria, was broadcast during the 2011-2015 period on local television stations (Canal 2 international, Ariane tv, Samba tv) and on a loop on local radio stations (FM 94, Crtv-centre, Magic FM, Royal FM). The artist-musicians who wrote this song played a major role in the process of stepping up communication to change behaviour, which is why they were real intermittent actors in the fight against malaria. In the same vein, another song entitled "Tons unis contre le paludisme" (Let's unite against malaria) was composed in 2010 by a group of local artists including X-Maleya, Duc-Z and Kate. This song was broadcast on a loop between 2011-2015 on CrtvFM94 and Crtv-Centre, urging people in Yaounde and the surrounding area to wear mosquito nets and sleep under them. The chorus of the song was particularly interesting, as it was an invitation to sleep under a mosquito net. It goes as follows:

"Sleeping under an impregnated mosquito net (impregnee)
Te protects against mosquito bites
Who gives malaria
Who gives malaria
Malaria must get out (eee

In short, with the implementation of the *"NightWatch"* and "K.O maludisme" programmes between 2010-2011, musical artists have become involved in the fight against malaria, within the framework of the CCC as defined by the PSNLP. The arrival on the scene of these intermittent actors has given a new impetus to the fight against malaria, as the messages conveyed by their songs have had a strong impact on households and the general public. In its sung version, the malaria anthem "K.O palu" is not ignored by any adult in the region. The use of renowned musical artists to raise awareness of malaria among the general public has been a major success.

[205] Source: kamerlyrics.net/lyrics-355-ko-palu-malaria-no-more-united-against-malaria.
[206] Ibid.

songs can also be lost as a seductive technique for persuading people.

In addition to musical artists, other intermittent actors have been involved in the field of malaria by communicating about the disease to the general public since 2002. These included politicians, intellectuals and health experts. The most emblematic politician in terms of media communication on malaria is the MP Marie-Rose NGUINI EFFA. Elie regularly spoke out in the media about malaria between 2014-2015. Elie was even an actress in an advertising spot that was widely broadcast in the audiovisual media in the social health field in Central Cameroon, which is why she was a major intermittent "actor" in the household awareness-raising process. The advert in question was entitled "K.O maludisme, tons unis contre le paludisme" and its message focused on "protecting children against malaria". The advert was broadcast en masse on Crtv tele, Canal 2, Samba TV, Ariane TV and on local radio stations Sweet FM, Magic FM, FM94, Crtv-Centre, Royal FM and Sky One Radio. People were "forced" to listen to the Honourable Marie-Rose NGUINI EFFA's message, as it was broadcast after each programme, practically every hour. In the same vein, the Executive Director of the NGO ISA, Olivia Ngou, regularly appeared on Yaounde radio stations to raise awareness of the risks of malaria. No fewer than fifteen radio programmes were produced by this social 'actor' between October 2015 and January 2016, half of which were broadcast on the regional station of crtv-centre[195] . A programme called "Family Health" broadcast on Royal FM constantly opened its doors to ISA's Executive Director. This "intermittent actor" also expresses his views on this pathology in programmes broadcast on television, notably on Canal 2 International[196] . Finally, it should be noted that several other radio and television programmes have invited health experts to talk about the issue of malaria in order to raise awareness among households as part of the communication campaign to change behaviour. The most emblematic of these programmes are:

-    **"Sant Secours"** is a three-minute micro-programme broadcast on Crtv radio since 2010 from Monday to Friday at 12.30 pm. The aim of the programme is to popularise public health issues, in particular the fight against malaria;

[195] Interview with Olivia Ngou, Cameroon Executive Director of the NGO malaria no more on 30 November 2020.
[196] This was the case with l' Nous chez Vous on 20 April 2020.

- **"S.O.S docteur"** is a programme broadcast on Crtv's national station every Saturday from 6.45am to 8am. The aim of the programme is to raise public awareness of the risk of malaria;
- **"Malaria in pregnant women"** is a special programme broadcast on Crtv radio on Wednesday 18 April 2018 between 5.30 pm and 5.55 pm in the run-up to the celebration of the World Malaria Day on 25 April 2018. The aim of the programme was to raise awareness among pregnant women of the risks associated with malaria;
- **"Sante plus actu"** is a programme broadcast on Crtv radio every Tuesday from 9.15am to 10am and presented by Marthe ADA MVONDO. On 22 April 2019, the theme of the programme was "acute malaria in children under the age of 05", with Dr Issa Ngosso as guest speaker. The aim was to raise parents' awareness of the risks of malaria in their children under the age of five.
- **"Family Health"** is a public health programme broadcast on the private radio station Royal FM. The programme constantly highlights the problem of malaria, emphasising the measures that can be taken to prevent it. Although broadcast in English, it attracts the participation of several listeners from Yaounde.
- **"PULSE"** which is a programme focusing on public health issues broadcast on Crtv-News. The issue of malaria is constantly raised. In a programme broadcast on 03 June 2018, the aim was to raise awareness of "new methods of combating malaria".
- **"Notre Sante" is a** public health programme designed to raise awareness of the risks of disease and how to avoid them. Broadcast every Saturday from 9am to 10am on Vision 4, it often focuses on the problem of malaria. One of its programmes focused on "preventing malaria in pregnant women and children".

By increasing the number of broadcasts and programmes about malaria, the media and the people who speak out are helping to step up the process of communicating about behavioural change and thereby raising awareness among households. The aim is to have a significant impact on the behaviour of *target groups* so as to get them involved in the fight against malaria through acts of prevention.

In short, the intensification of the modern fight against malaria from 2022 onwards will be reflected empirically by an increase in communication aimed at changing people's behaviour by both collective and individual actors. Collective players such as the state and NGOs are mobilising instruments and resources to have a significant impact on people's behaviour. They are followed in this by individual actors such as grassroots political and administrative elites and so-called intermittent actors. Within this process of large-scale communication, the state retains a monopoly in terms of the instruments of public action it mobilises and the extent of its sphere of action in both urban and rural areas. The intensification of the modern fight against malaria in the geographical area of Central Cameroon is also reflected in a very beautiful way by the rise in power of the public authorities in the fight against the disease.

Paragraph 2: The rise of public authorities in the field of malaria

The intensification of the modern fight against malaria in the geographical area of Centre-Cameroon is illustrated by the growing importance of public authorities in the field of the disease from 2010 onwards. Public authorities are public actors, i.e. those invested with the authority of the State and public power. They are referred to in this way in contrast to private actors, who are also involved in the process of implementing public policy. To curb the threat of malaria to a considerable extent, the state authorities are investing in the field of malaria by stepping up their intervention in accordance with the principle of "multisectorality".

In practical terms, the growing power of the public authorities in the field of malaria is reflected in the taking of several "strong" public decisions designed to "regulate" the fight against malaria in health facilities (FOSA). The decision occupies an essential place in the hierarchy of public acts. For it is at this precise moment that the game is played out. A politician makes a decision. Solutions are chosen. A series of events ensues. Those affected by the public policy in question will see their situation affected. The future seems irreversibly committed[197] . The decision thus appears to be the moment when public actors actually take action, making optimal choices that are likely to have an impact on people's lives.

Between 2010 and 2015, the fight against the malaria problem in the Centre region gained new momentum with the adoption and implementation of a number of public decisions. The actors involved are essentially public authorities. They are mobilising "regulatory" instruments to take action in the social field. From that moment on, malaria was associated with a real "public rationality" and became a *de facto* "public problem". The fight against malaria has undergone a real change of direction, as it no longer has anything to do with the unorthodox, archaic and traditional methods used before the adoption of the *"Roll Back Malaria"* programme. It now bears the stamp of "modemity". It is becoming even tougher with the rise in power of both the central public authorities (A) and the local public authorities (B).

A- The rise of central public authorities in the field of malaria

The growing power of central public authorities in the field of malaria is reflected in a number of public decisions taken between 2010 and 2014. Analysing the decisions taken by central public players means studying the 'central decision-making environment'[198] or, to put it another way, 'the first circle of decision-making', i.e. the circle through which all major decisions pass[199] . In France, as in Cameroon, it is made up of the President of the Republic and the Prime Minister. The decisions taken by these central public authorities have a real impact on all the sub-systems of the State. In fact, all decisions taken at national level have a considerable impact at local level. Alain FAURE illustrates this reality well when

---

[197] See MENY (Y) and THOENIG (J-C), Op. cit... p. 189.
[198] GREMION (C), quoted by MULLER (P), Les politiques publiques. Op. cit... p. 39.
[199] MULLER (P), Idem. p. 40.

he writes that "local public action is to some extent *contained in* public decisions which are conceived and set out within central administrations"[200] . As a result, the decisions taken by Cameroon's central public players have a real impact at all local levels, including in the Centre region.

Decisions taken by central public authorities are a social activity that is "rational in purpose"[201] because their aim is to solve a given social problem. In the context of the fight against malaria, the central public authorities are represented by the President of the Republic and the Minister of Public Health. These two public actors have taken important decisions since 2010 to combat the problem of malaria in Cameroon. These decisions are not without effect in the Centre-Cameroon region. The central public authorities are intervening in the field of malaria by issuing what Max Weber calls "administrative regulations"[202] . They thus mobilise "regulatory resources" to act on the social fabric. The President of the Republic (1) and the Minister of Public Health **(2) are the** ones involved in stepping up the fight against malaria in the Centre-Cameroon region.

1. The rise of the President of the Republic in the field of malaria

In Cameroon, the President of the Republic is the Head of State, elected by the entire nation, and defines the nation's policy[203] . At the highest level, he is the "central decision-making body", i.e. it is through him that all the important decisions of the State pass. The President of the Republic of Cameroon, who has long been involved in the field of malaria on a residual basis, stepped up his role in the fight against the disease from 31 December 2010 by taking 'strong' and 'innovative' decisions aimed at limiting the spread of the disease. In doing so, he is stepping up the fight against malaria.

Between 2010 and 2014, the President of the Republic stepped up the fight against malaria in the Centre region. This public authority intervenes in the field of malaria using the "representative democratic feedback" mechanism[204] defined by Yves MENY and Jean-Claude THOENIG. From this perspective, the needs and problems of local residents are seen as imbalances which, as they grow in scale, are dealt with through procedures and by qualified intermediaries, political intermediaries who, on behalf of their constituents, put pressure on the relevant public authority to intervene[205] . In other words, the rise to power of the President of the Republic in the field of malaria "resembles a process of   escalation based on isolated requests"[206] . It is, in fact, the day-to-day experience of the

---

[200] Read FAURE (A), <u>La question territoriale. Pouvoirs locaux. action publique et politique Is)</u>. Political science. Universite Pierre Mendes-France - Grenoble II, 2002. p. 44.

[201] Read WEBER (M), <u>Economy and Society 1. The Categories of Sociology</u>. Agora les Classiques, Pocket, Pion, 1995. p. 55.

[202] These are all rules which apply to the behaviour of the administrative management as well as, according to an expression which has become commonplace, to that of the members "towards the group". See WEBER (M), <u>Economie et societe</u>. Op. cit... p. 93.

[203] See article 5 (1) of law n°96/06 of 18 January 1996 revising the constitution of 02 June 1972.

[204] Read MENY (Y) and THOENIG (J-C), <u>Politiques publiques.</u> Op. cit... p. 161.

[205] Ibid. p. 161.

[206] Ibid. p. 161.

population that has prompted the President of the Republic to   take   "strong" decisions on malaria.   "decisions relating to

malaria on the basis of a "bottom-up rationality". These so-called "strong" decisions are of two kinds: those favouring the fight against malaria in children under the age of five and those favouring the fight against malaria in people over the age of five.

With regard to measures to combat malaria in children under five years of age, it should be noted that the

On 31 December 2010, the President of the Republic, Paul BIYA, surprised health workers and nationals of the Central Cameroon region by making the following statement: "I *am pleased to announce that I have decided to abolish the cost of treating simple malaria in children under the age of five"*[207] . With these words, the President of the Republic has, in his own way, stepped up the fight against malaria. As children under five are the worst affected by the disease, the Head of State has opted for free treatment for this segment of the population in the event of an episode of uncomplicated malaria. In the aftermath of this decision, it could be heard in the streets of Yaounde that "Paul Biya is offering 600 francs to sick children", as reported by the daily *Le messager in* its 05 January 2011 issue.

The decision taken by the President of the Republic has been dismissed by the public as a "breath of relief", because according to doctors, *"eighty per cent of cases of malaria in children under the age of five declared in hospitals are in the simple form"*[208] . In reality, this means a reduction in the financial burden on parents, who will no longer have to pay a single penny to ensure the recovery of their sick children. In the past, parents had to pay between 1,000 and 2,000 CFA francs for treatment, but now it's free of charge, to ensure a better response to malaria in children under the age of five.

In May 2011, as part of the fight against malaria in children under the age of five, the President of the Republic, Paul BIYA, decided to make the Rapid Diagnostic Test (RDT)[209] free of charge for this social category[210] . Previously chargeable, the paracheck-pf RDT for children will be officially free from 2011, giving families

---

[207] Excerpts from a speech given on 31 December 2010 during the traditional ritual of presenting the vows to the nation.

[208] Interview with Mr MENDIMI NKODO Joseph Marie, Director of the Jamot Hospital in Yaounde on 04 January 2021.

[209] The State of Cameroon opted to introduce the RDT in 2008. Its principle is based on the detection of *plasmodium sp* antigen or antibodies directed against plasmodium (anti-plasmodium antibodies) in the blood (whole blood or plasma) of the test subject. The vast majority of RDTs are read within the first twenty minutes after they are performed; the results can be stored for three days at 2-8°C. Depending on the age of the patient to be tested, a distinction is made between the Diaspot RDT and the paracheck-pf RDT. The Diaspot RDT tests patients of all ages, while the paracheck-pf RDT tests only children under the age of five. See DJEUTCHOUANG SAYANG (G), Interet de l'utilisation des tests de diagnostic rapide du paludisme sur le cout et l'efficacite de la prise en charge des patients febriles a Yaounde Cameroun, Op. cit... p. 93.

[210] See the MINSANTE information note on the fight against malaria of 18 April 2016.

[223] Interview with Ms ONANENA Bernadette, a national of the Ntui district, on 16 June 2021.

[224] See the MINSANTE information note on the fight against malaria dated 8 April 2016.

[225] Read MENY (Y) and THOENIG (J-C), Politiques publiques. Op. cit. P. 161.

more hope of "saving" their sick children under the age of five. The Head of State's logic is quite clear: since the treatment of uncomplicated malaria is now free for children under five, the detection of this case of malaria should also be free. In this way, parents can officially test and treat their sick children free of charge. But there was also the question of how to treat severe malaria in children under the age of five. In fact, after deciding that treatment of uncomplicated malaria is now free for children under the age of five, the people of the Centre were faced with a problem. In hospitals, cases of malaria in children are increasingly being declared "serious", and in such cases, people have to pay. In an interview, a woman from the Ntui health district explained the situation to us:

"When the Head of State decided that treatment of uncomplicated malaria was free for children under the age of five, every time I took my son, who was in that age group, to hospital, the doctor would always say that he had severe malaria, and at that point you had to pay the treatment costs, which sometimes amounted to five thousand francs in the event of complications. It was as if the decision to make simple malaria free had made this form of malaria disappear"[223] .

In June 2014, the President of the Republic decided that the treatment of severe malaria would now be free for children under the age of five[224] . As a result, children in this age group are completely exempt from paying any costs in the event of malaria, whether simple or severe. This decision means that the difficulties encountered by the population in relation to malaria reach the "central decision-making environment" through a process of "representative democratic escalation"[225] following a bottom-up logic.

In short, it can be said that the decisions taken by the Head of State proclaiming the free treatment of malaria and the RDT byacheck-pf - are a step in the right direction.

in children under the age of five are marked by "political will" in the sense of Haroun JAMOUS[211] . His aim was to give a new impetus to the process of stepping up the fight against malaria among the most vulnerable sections of the population.

With regard to people over the age of five, it should be noted that in February 2011, the President of the Republic relaxed the cost of treating simple malaria in this category of the population. Where there used to be arbitrariness, since February 2011 the cost of treating an episode of uncomplicated malaria for the over-fives has been set at between 200 and 250 CFA francs[212] . This is another powerful measure that 'minimises' the cost of treating malaria, this time in adults, by putting an end to the arbitrary nature of health facilities. In the opinion of some people from the Centre region, "if *you talk about 250 francs, you're still getting free hospital treatment*"[213] . The fight against malaria has thus officially made significant progress with the relaxation of the cost of treatment for uncomplicated malaria. In the same vein, in March 2012, the cost of RDT for people over the age of five was

---

[211] Read JAMOUS (H), Sociologie de la decision. La reforme des etudes medicales et les structures hospitalieres, Paris, CNRS, 1969, p. 175.
[212] See the MINSANTE information note on the fight against malaria of 18 April 2016.
[213] Interview with M. ONANENA Bernadette, 16 June 2021.

set at 200 CFA francs by the President of the Republic[214] . This is another "strong" measure to ease the financial burden on the population. Elies can now be tested and treated for less than 500 CFA francs.

The decisions taken by the Head of State have been widely echoed in the Centre region. In this region, people could be heard expressing a certain satisfaction, because malaria no longer has a cost as such. The President of the Republic is thus building a new era in the process of institutionalising the fight against malaria. The disease has truly acquired the status of a "public problem", because it is now on the agenda of the highest State authority, and is the subject of one or more public decisions[215] . The fight against malaria is now based on a real "public rationality", especially as MINSANTE has been involved in the "toughening up" process since 23 June 2014 through another "strong" decision regulating the fight against malaria in health facilities.

2. The rise of the Minister for Public Health in the field of malaria

In the fight against epidemics, the "central decision-making environment" is not the exclusive preserve of the Head of State, who interacts with other central public players in the process of building the public response to malaria. In fact, the President of the Republic is not the only central public authority to have stepped up its involvement in the field of malaria through "strong public decisions" since the adoption of the *"Roll Back Malaria"* programme. In fact, it could not be otherwise, since heuristically, "the decision takes the form of a continuous flow of decisions and one-off arrangements, taken at different levels of the action system, which must be analysed as a set of decision-making processes"[216] . Sequential analysis clearly shows that there is no single decision; there is no one decision which, at a clearly identifiable moment, determines the content of a public policy[217] . In other words, the decision is a process in which several public players are involved.

As part of the fight against malaria in Cameroon in general, and in the Centre region in particular, President Paul Biya is not the only person to take decisions governing the treatment of malaria in hospitals. Although the Head of State appeared to be the only public player in the decision-making process between 2010 and 2014, he did not monopolise the process because from 2014 onwards, MINSANTE entered the field of malaria by taking a "strong" decision. In fact, MINSANTE is intervening to complement the decisions of the President of the Republic, in a kind of "solidarity of the first decision-making circle". MINSANTE is therefore following the path set out by the Head of State, by providing more details on the ways in which malaria is treated in hospitals. It has done so by means of a decision dated 23 June 2014[233] , article 9 of which sets the price of treatment for severe malaria.

---

[214] See the MINSANTE information note on the fight against malaria of 18 April 2016.
[215] See GARRAUD (P), "Agenda/Emergence", in BOUSSAGUET (L), JACQUOT (S) and RAVINET (P) (eds.), Dictionnaire des politiques publiques. Op. cit... p. 55
[216] See MULLER (P) and MENY (Y), L'analyse des politiques publiques. Paris, Montchrestion, 1998, p. 103.
[217] See HASSENTEUFEL (P), Sociologie politique : l'action publique. Loc. cit... p. 121.

In June 2014, in line with the President of the Republic's intention to step up the fight against malaria, MINSANTE stepped up its activities in the field of malaria with a decision. This was decision n°0406/D/MINSANTE/CAB/ of 23 June 2014 on the pricing of treatment for severe malaria in Cameroon[218] . Article 9 of the said decision clearly states that: "the prices for the transfer of treatment within the **framework of inter-generation solidarity for an episode of severe malaria** at health facility level are established as follows:

- Children under five (excluding hospitalisation costs and any complications): **free of charge;**
- Pregnant women from the second trimester of pregnancy (excluding hospitalisation costs and any complications): **four thousand (4,000) CFA francs;**
- Patients over five years of age (excluding the cost of hospitalisation and any complications): **eight thousand (8,000) CFA francs"**[219] .

Basically, the MINSANTE decision of June 2014 regulates the prices of severe malaria treatment in hospitals for all types of patients. The decision puts an end to the "arbitrary reign" of doctors, who can no longer set the price of treatment for severe malaria as they see fit, because a legal framework now exists. It should be stressed, however, that article 9 of the June 2014 decision does not apply to possible complications due to severe malaria or to hospitalisation costs, which remain payable!

The day after its publication, the MINSANTE decision was implemented in hospitals in the geographical area of Central Cameroon, although there were several distortions here and there. In fact, a distinction must be made between the implementation of the decision of 23 June 2014 in Yaounde hospitals and in hospitals in the periphery.

This will enable us to highlight 'the production of local order systems for public action'[220] . In other words, public policy is not implemented uniformly across a territory, because each sub-system has its own realities which territorialise the process of applying decisions. Thus, a decision will not be implemented in the same way in space A as in space B.

In the Yaounde hospitals, the decision of 23 June 2014 was strictly implemented, i.e. taking into account both the spirit and the letter of the decision. In this case, the problem with the process of implementing the Minister's decision was that it was not properly understood by the local population. It could even be argued that the decision taken by MINSANTE was not explained in detail to the public. As a result, some patients considered doctors to be zealots who went against MINSANTE's directives in order to make *"gombo"*, as the local jargon frequently says. On this particular point, the director of the Jamot hospital made the following clarifications:

---

[233] The full decision can be found in appendix 3 of this report.
[219] Read article 9 of the aforementioned ministerial decision.
[220] Read LASCOUMES (P) and LE GALES (P), Sociologie de 1 action publique. Armand Colin, 2ᵉ ed. 2012, p. 38.

"Free care for children under five is now available in health facilities. Those who say that we are not complying with this requirement misunderstand the Minister's measure. She's talking about children under five, but parents often come with children aged five plus a few months and want to take advantage of the measure too. What's more, there are often complications that have to be paid for"[221] .

The difficulty in implementing the Minister's decision stems from the fact that, according to the director of the Jamot hospital, the sick people who go to the hospital misunderstand Article 9 of the decision. Thus, according to this testimony, the parents of sick children did not understand what was meant by "children under five" and refused to accept that complication cases had to be paid for. It would therefore have been necessary to educate them beforehand, or even better, to raise their awareness, to avoid statements like these

"we're still paying for severe malaria in children under five in hospitals"[222] .

In the peripheral health facilities (FOSA), the decision of 23 June 2014 and its article 9 were not strictly applied the day after they were adopted. Here, treatment of simple or severe malaria for children under the age of five continued to be paid for. This social reality was encouraged by the fact that the vast majority of the population in rural areas is illiterate and could therefore be "misled" by certain crooked doctors. This was the case at the Ntui district hospital, where treatment for severe malaria in children under the age of five continued to be charged for after 23 June 2014, as did the RDT. A local woman from the town of Ntui gave us the following account in an interview: "*Every time I went to the hospital with my little daughter, I was always asked to pay, and I never treated her for free, whether in 2014, 2015 or 2016. And even the malaria test has always cost money here, it costs exactly one thousand francs FCFA*"[223] . The same reality has been experienced by the people of the Yoko health district, if the testimonies of many families are anything to go by[224] . Thus, in the immediate aftermath of its adoption, the decision of 23 June 2014 proved difficult to implement.

In short, the rise of the public authorities in the field of malaria has revolutionised the fight against the disease. The decisions taken by these central public players (the President of the Republic and MINSANTE) have been decisive in the process of stepping up the fight against malaria in the Centre region, despite the distortions observed during their implementation in the social field. The Elies regulate the treatment of malaria in health facilities. Treatment of malaria in children under the age of five is now officially free. The situation of pregnant women suffering from malaria has also been officially improved. Insofar as they are aimed at reforming the health administration in the fight against malaria and improving the state of health of its citizens, the decisions taken by the President of the Republic and

---

[221] Interview with Mr MENDIMI NKODO Joseph Marie, Director of the Jamot Hospital in Yaounde on 4 January
[234] Ibid.
2021.
[222] Recurrent discourse among parents of sick children in Yaounde in the wake of the 23 June 2014 decision.
[223] Interview with Ms Onanena Bernadette, a native of the town of Ntui, on 22 April 2021.
[224] A field survey was carried out among six households in the "Mengang" and "Yoko" neighbourhoods on 4 and 5 May 2021.

MINSANTE cannot be described as "parade decisions"[2]' but rather as "voluntarist decisions", i.e. decisions marked by a real "political will". In this respect, they gave a new impetus to the process of institutionalising the modern fight against the malaria problem, by regulating the fight against malaria in hospitals. Following the intervention of the central public authorities, the local public authorities in the Centre region will enter the process of stepping up the fight against malaria between the third quarter of 2014 and the first quarter of 2015, through more concrete measures in the social field.

8- The rise of local public authorities in the field of malaria: the launch of "hygiene and sanitation" campaigns

After the intervention of the central public authorities, the local public authorities are stepping up their role in the process of stepping up the fight against malaria through public decision-making. Analysing the decisions taken by local public authorities still means paying some attention to the actors involved in "peripheral power" '[212] or the "second decision-making circle" '[213] . Local public players are represented here by the political and administrative authorities (mayor, sub-prefect, prefect, governor). By virtue of their proximity to the population, they are the architects of the 'territorialisation of public action', which is 'the dimension most immediately associated with the rhetoric of proximity'[225 226 227 228] . Indeed, bringing public decision-making closer to the places where social problems emerge and are resolved is a central element in the legitimisation, through proximity, of local public health policies.

The local public authorities are "basic political and administrative agents" who intervene in the process of fighting the malaria problem as a complement to State action. Since the third quarter of 2014, they have been taking "strong" decisions designed to give a new dimension to the fight against malaria in the social and health field in Centre-Cameroon. In this way, they are making the fight against malaria a reality "from below", as they are sociologically closer to the people than the central public players mentioned above.

The local authorities are institutionalising the modern fight against malaria between 2014-2015 with the official launch and implementation of the "hygiene and sanitation" campaign. This campaign is in line with MINAT's circular letter n°0040/LC/MINAT/DCTD of 04 April 2000, which institutes the organisation of public health and hygiene campaigns in all localities in Cameroon. In the Centre region, this campaign was officially launched by Governor OTTO Joseph Wilson in the third quarter of 2014 under the slogan *small no be sick"*. Its aim is to

---

[225] Read JAMOUS (H), Op. cit... p. 175.

[226] This expression, borrowed from Pierre GREMION, is understood in the context of this study as the decision-making power of peripheral or local players who, by virtue of their prerogatives, can also "regulate" the fight against malaria in their localities.

[227] Expression borrowed from Pierre MULLER and understood in the context of this study as the area of decision-making reserved for sectoral public players as opposed to central public players.

[228] See BERTHET (T), "L'Etat social a l'epreuve de l'action territoriale. Postmodemite et politiques publiques de proximite dans le champ de la relation emploi-formation", in FAURE (A) et NEGRIER (E) (dir), Les politiques publiques a l'epreuve de l'action locale : critique de la territorialisation. Paris, L'Harmattan, 2007, p. 47.

promote a better response to so-called dirty diseases, of which malaria is the most prominent. Between 2014 and 2015, the hygiene and sanitation campaign was implemented by local authorities in Yaounde **(1)** and in rural areas **(2).**

1. Hygiene and health campaigns launched in Yaounde

In August 2014, the prefect of the Mfoundi department, Mr Jean-Claude TSILA, launched the "hygiene and cleanliness" campaign in the city of Yaounde by simple recommendation. This is a competition aimed at the administrative structures of the Mfoundi department, at the end of which the cleanest commune in the city of Yaounde will be selected. The ultimate aim of the competition was to combat diseases caused by dirt, such as malaria. In the seven communes of the Mfoundi department, from August 2014 onwards, there was a strong deployment of people and households in the field of cleanliness, aided by the company known as "Hygiene et Salubrite du Cameroun" (HYSACAM).

In the commune of Yaounde six, the Mayor, Mr Paul-Martin LOLO, launched the "clean commune" operation, which is the territorialisation of the "hygiene and sanitation" campaign in this area. He proceeded with the massive recruitment of two hundred young pupils and students as part of a holiday course to clean up the commune, more specifically the Biyem-Assi district. From Monday to Friday over a two-week period, these young people were sent out into the field, each wearing a blue smock, a badge stamped "Yaounde 6 Commune" and green boots. Ten skips were made available to them to help them get around the commune. The work of these young people consisted of:

- Encourage people to clean up around their homes, especially toilets when they are out of date. This operation was made possible by raising awareness among householders;
- Collect household rubbish in the vicinity of shopping centres;
- Sweep the public highway every morning and clear the grass.

Through the "clean municipality" operation, Mayor Paul-Martin LOLO aimed to make the municipality of Yaounde six "the cleanest municipality in Cameroon"[229] . During the operation, households and market vendors were also very active. Strongly sensitised by the municipal workers, they were the real artisans of cleanliness and hygiene in the Biyem-Assi district. Every Saturday morning, the men could be seen clearing the grass around their houses and cleaning out the small puddles and pools of water. At the place known as "Grand Marche de Biyem-Assi", the vendors cleaned the ponds, shops and trading posts every morning. They also swept the trading areas and collected rubbish.

The most insalubrious area of the Biyem-Assi district, known as the "nage-glisse"[230] , which is conducive to the large-scale spread of mosquitoes, has been the

---

[229] According to Alain Mendouga, a former trainee at Yaounde 6 town hall, in an interview on 19 September 2020.
[230] The expression "nage-glisse" is an indexical expression with a purely local meaning. Elie refers to the area of the Biyem-Assi district as the most unhealthy because of the presence of untended low-lying wetlands. People refer to it as a "swim-slide" because of its configuration.

focus of a new study.

The area was treated and landscaped by local authority and HYSACAM staff from August 2014. As part of the "clean municipality" operation, this marshy area was cleared and levelled using motorised and sophisticated equipment. The run-off and stagnant water in the area has been drained, limiting the appearance of mosquitoes. A local resident testified to this during an interview:

"It was Wednesday 16 April 2014 when agents from the commune and hysacam arrived in the nage-glisse sector with skips, peles and some motorised equipment. They proceeded to clear and level the marecage. I remember that evening, my family and I slept very well, not a single whistle from [247] mosquitoes".

The *"small no be sick"* campaign and its local "clean commune" counterpart have enabled the local authorities and the people of Biyem-Assi to reduce the risk of malaria, not only by keeping the commune clean, but also by drastically reducing the number of female anopheles, the vector par excellence of plasmodium.

In the commune of Yaounde 1er , the mayor, Mr Emile ANDZE ANDZE, launched the "hygiene and sanitation for all" (HST) campaign in August 2014, a slogan linked to the territorialisation of the regional *"hygiene and sanitation"* campaign. The campaign was essentially based on raising people's awareness through a team of young people recruited by Yaounde town hall (^{er}). For the mayor, the aim was to encourage people to promote hygiene measures as much as possible in their immediate environment. The Tongolo, Etoa-Meki, Nylon, Manguier and Elig-Edzoa neighbourhoods were the main targets of the campaign, as they are reputed to be the most insalubrious in the commune. Local residents were required to clean the toilets, most of which were unfinished and shared, on pain of a fine. As a result, the people are obliged to refurbish the tin toilets, otherwise they will be destroyed by the Yaounde Urban Community. This is an accident-prone area, because not only is it slippery, but it can also lead to swimming for those who find themselves inside.

[247] Interview with Mr Dieudonne Nana, aged 58, resident of the Biyem-Assi neighbourhood and resident of the "nage-glisse" area. Interview conducted on 4 September 2020.

This was the case in almost all the neighbourhoods of the Yaounde 1er district, with the exception of the Etoa-Meki and Tongolo neighbourhoods, where unfinished communal toilets were destroyed by Yaounde Urban Community (CUY) agents. What we observed was that in the commune of Yaounde 1er , the CUY had a strong presence and was keeping a close eye on things. In this area, there was indeed a juxtaposition of two actors in charge of implementing the "hygiene and sanitation" campaign, namely the commune of Yaounde 1er and the CUY. The former was involved in awareness-raising and prevention, the latter in repression. Jack Bauer"[231] destroyed "seventy unplumbed toilets between August 2014 and January

---

[231] It's a nickname given to the delegate of the government of the Yaounde urban community by the local population because of his ability to easily carry out the repressive phase of the hygiene and sanitation policy by destroying toilets, for example. In reality, Jack Bauer is a very popular character in the American series "24 heures

2015 in the Tongolo and Etoa-Meki neighbourhoods"[232] .

The "hygiene and cleanliness for all" campaign also invited shopkeepers in the Etoudi and Elig-Edzoa markets to demonstrate their responsibility by collecting rubbish and sweeping the streets. In the same vein, the commune of Yaounde[er] , through its agents, set up rubbish bins along the main roads and put up posters reading "No urinating here". There were 115 rubbish bins in the district of Yaounde 1[er] , the largest of which were positioned in the Etoudi district, where the Palais de l'Unite is located.

In short, the "hygiene and cleanliness" campaign officially launched by the Prefect of Mfoundi in August 2014, has been reappropriated and territorialised in the different districts of Yaounde. Thus, in Yaounde 1[er] we spoke of "hygiene and salubrity for tons", in Yaounde 2[e] we spoke of "clean city operation", in Yaounde 3[e] we spoke of "sanitation campaign", in Yaounde 4[e] there was talk of a "fight against insalubrities", in Yaounde 5[e] there was talk of a "clean-up operation", in Yaounde 6[e] there was talk of a "clean municipality" and in Yaounde 7[e] there was talk of a "major clean-up campaign". In other words, the administrative and municipal authorities of Yaounde's seven districts followed the slogan issued by the regional governor and the Prefect of Mfoundi. The aim of the *"small no be sick"* campaign was to create a competitive space between Yaounde's different arrondissements, at the end of which the cleanest and healthiest commune would be elected. Practically similar actions were undertaken in the various sub-systems of Yaounde by the local political and administrative players, i.e. the municipal and administrative authorities: recruitment of young people for awareness-raising operations, mobilisation of municipal staff for the maintenance of municipal roads, signature of special contracts with HYSACAM, introduction of penalties for users who dump household waste in the middle of the urban centre and outside the rubbish bins, awareness-raising campaigns in areas open to the public such as markets, repression of recalcitrant behaviour by the CUY, etc. In short, in the city of Yaounde, the implementation of the "hygiene and sanitation" operation during the third quarter of 2014 really cleaned up the city and, as a result, protected many families from the mosquitoes and unsanitary conditions that cause malaria. The same action has been taken in outlying areas.

2. The launch of "hygiene and health" campaigns in rural areas

The arrondissements of the nine other departments in the Centre region, considered here as rural areas compared to Yaounde, have also implemented the "hygiene and sanitation" campaign. In the department of Mbam-et-Kim, the prefect, Mr MAMOUDOU, officially launched the campaign on 15 August 2014. During the campaign, people in the area were urged to comply strictly with hygiene and sanitation measures. In the commune of Ntui, the mayor, Mr Jerome ONDOBO MONO, himself carried out a large-scale awareness-raising campaign among local

chrono".
[232] Interview with an official from Yaounde town hall, who preferred to remain anonymous.

residents at a place called "centre". The Ntui market was the first site to be used for the hygiene and sanitation campaign in the town. Twenty-five (25) rubbish bins were made available to vendors for the dumping of rubbish from commercial activities. Pistachio leaves and cassava sticks were to be dumped in these bins as a matter of priority. Traditional sweepers have also been made available to vendors to keep the counters and commercial ponds clean at all times. The public toilets inside the Ntui market should be treated appropriately by the local authority staff.

A team of young people, recruited as part of the usual holiday training programme, was sent out to various neighbourhoods to clean up the area around the water wells, which are also densely populated areas in Ntui. As the town is not covered by the Camwater network, the entire population heads for these wells, which are the only reservoirs of drinking water in the town of Ntui. Clearing the grass and treating the stagnant water around the wells was the 'job description' of the young people on holiday training at the town hall. What's more, the hygiene and sanitation campaign in Ntui also meant that households were heavily involved in the process. In fact, every household was required by the local officials to clean up the area around their house or face financial penalties (payment of taxes). The maintenance of communal toilets in the neighbourhoods became a compulsory operation.

In the Yoko health district, the mayor, Mr Dieudonne ANNIR TINA, launched the "hygiene and cleanliness" campaign in the area on 22 August 2014. *"As its name suggests, the aim of this campaign is to promote hygiene and cleanliness in the commune. Through it, we were able to protect many families from certain diseases"*[233] . The fight against disease was thus the pretext for launching the "hygiene and sanitation" campaign in the commune of Yoko. In the "prison" district, the many water horns littering the alleyways and houses were treated specifically to limit the spread of mosquitoes. In the "Yoko" district, the many grassy areas there have been cleared by local residents. The "Yoko market" was the priority area for the hygiene and sanitation campaign. The large "hangar" there was specifically cleaned up by the butchers and other vegetable sellers.

Cleaning latrines was the focal point of the hygiene and sanitation campaign in Yoko. In this area, *"ninety per cent of households use communal toilets"*[234] . These toilets are *"a major health risk for the local population, which is why the 2014 hygiene and sanitation campaign focused on them"*[235] . Local people were therefore urged by local authority officials to fit out their toilets. During the campaign, straw toilets were replaced by mud toilets, and the floor had to be made of cement. The grass around the latrines must be cut. It is clear from this that the "hygiene and sanitation campaign" in Yoko was first and foremost a campaign to combat the so-called "dirty diseases", of which malaria is a major one.

In the department of Nyong-et-Mfoumou, the prefect, Mr Charles KAMGA,

---

[233] Interview with Mr Dieudonne ANNIR TINA, Mayor of the commune of Yoko, on 7 November 2020.
[234] Interview with Mr Dieudonne ANNIR TINA, Mayor of the commune of Yoko, on 7 November 2020.
[235] Ibid.

launched the "major hygiene and cleanliness campaign" on 22 August 2014. As in all the outlying departments, this campaign aims to promote hygiene and cleanliness in all the arrondissements of the department in order to protect households from so-called dirty diseases. In the Akonolinga district, the mayor, Mr Joseph ESSAMA, is organising the hygiene and sanitation campaign on a territorial basis, with the introduction of special measures. Teams have been set up to sweep the main roads and crossroads. They are made up mainly of young pupils and students who are doing a holiday course at Akonolinga town hall while waiting for the start of the new school and academic year. The area known as the "administrative centre" was a priority zone for sweeping and cleaning the gullies. Mobile teams of local authority staff toured the neighbourhoods to raise awareness of the need to keep latrines clean. Penalties were imposed on those who failed to keep their toilets clean.

In short, the "small no be sick" campaign launched by the governor of the Centre region in the third quarter of 2014 is part of the drive to step up the modern fight against malaria. The campaign was rolled out across the region's different sub-systems by grassroots political and administrative players and local people. Whether in Yaounde or on the outskirts, the administrative, municipal and legislative authorities were the main architects of the implementation of this hygiene and sanitation campaign in their areas. The aim of the campaign was to ensure cleanliness in the various communes of the Centre region, in order to protect the population from a number of diseases, including malaria.

The rise in power of local public authorities augurs a new dynamic in the process of intensifying the fight against malaria, while at the same time institutionalising the modern fight. Being closer to the local populations than the central public players, the local public authorities are translating into action what Thierry BERTHET calls "the rhetoric of proximity"[236] which consists of players dealing with a social problem from its place of emergence. Inadequate sanitation in towns and cities is the work of the local population, and to combat it, local public action is needed to give the fight against malaria a local flavour. Such is the quintessence of the intervention of local public authorities in the fight against insalubrities in the Centre region during the third quarter of 2014 and the first quarter of 2015.

In short, the intensification of the modern fight against malaria in the geographical area of Centre-Cameroon is based on the accentuation of communication to change behaviour and the increased involvement of the public authorities in the process. These appear to be the most convenient and effective means of "countering" the threat of malaria. From 2002 onwards, a vast household awareness campaign was launched in the region. This mobilised a multitude of players to translate the Behaviour Change Campaign (BCC) into concrete action. Collective actors, such as

---

[236] See BERTHET (T), "L'Etat social a l'epreuve de l'action territoriale. Postmodemite et politiques publiques de proximite dans le champ de la relation emploi-formation", in FAURE (A) et NEGRIER (E) (dir), <u>Les politiques publiques a l'epreuve de l'action locale : critique de la territorialisation</u>. Paris, L'Harmattan, 2007, p. 47.

the State and NGOs, are mobilising public action tools and resources to raise awareness among households about the benefits of using a mosquito net and complying with hygiene measures. They are supported in this by individual actors, such as political and administrative elites and media personalities. The CCC is an operation that obeys a "multi-actor" dynamic in which the State retains a monopoly.

The *"Roll Back Malaria"* programme, adopted in 2002, is based on the principle of "multisectorality". This principle means that the fight against malaria is no longer a matter for health professionals, but for society as a whole, including the public authorities. In this way, the fight against malaria is becoming desectoralized through the involvement of several players who were previously outside the dynamic of the fight against malaria. In addition to the state, NGOs, political and administrative elites and intermittent players, the public authorities have been heavily involved in the process of stepping up the fight against malaria since 2010. These authorities are both central and local, and intervene in the field of the disease through a number of public decisions that will be implemented as best they can in the geographical area of Centre-Cameroon.

The intensification of the fight against malaria from 2002 onwards is not only based on increased communication to change people's behaviour and on the public authorities stepping up their efforts. It also saw the organisation of ITN/MILDA distribution campaigns.

SECTION II

The 5y51ётаИ5аИоп free distribution of mosquito nets

The intensification of the fight against malaria since the adoption of the *"Roll Back Malaria"* programme in 2002 has also been reflected in the field by the provision to the population of a tool to help them fight back against this disease. The first free distribution campaigns began in 2003. The distribution of ITNs/MILDAs is a multi-factorial process, involving a multitude of public and private players. For private actors, the distribution of Mil is achieved through the mobilisation of "material resources"[237] and "knowledge"[238] ; for public actors and the State, it is achieved through the mobilisation of incentive and communication tools[239] . Within the net distribution process, the state has a monopoly and is the dominant actor in the process.

---

[237]This category of resources is essentially collective. It includes the financial resources (budget), human resources (available personnel) and operational resources (premises, logistical and IT resources, technical tools, etc.) available to public or private players. See KNOEPFEL (P), LARRUE (C), VARONE (F) and HILL (M), "Policy resources", in <u>Public Policy Analysis</u>, Op. cit... pp. 63-88.

[238] They refer both to the information and knowledge available to a player and to his ability to interpret, trailer and integrate them into public action strategies. See KNOEPFEL (P), LARRUE (C), VARONE (F) and HILL (M), Op. cit... pp. 63-88.

[239] In their work, Pierre LASCOUMES and Patrick LE GALES distinguish five types of instrument on the basis of which the action of the State and public players is materialised in the field, namely: "Legislative and regulatory"; "Economic and fiscal"; "Conventional and incentive"; "Informative and communicational"; "Norms and best practice standards". See LASCOUMES (P) and LE GALES (P), <u>Gouverner par les instruments</u>, Paris, Presses de Sciences Po, 2004, p. 351.

The free mass distribution of ITNs/MILDAs to the population is the implementation of strategic priority 2 of the PSNLP devoted to vector control. It is now indisputable that the female anopheles is the vector par excellence of malaria transmission through the plasmodium it inoculates into the bodies of healthy people. Vector control is becoming the surest way of avoiding any risk of malaria through the mass distribution of LLINs to households. For the many players involved in this process, the distribution of mosquito nets to households appears to be a decisive operation in curbing the threat of malaria. It thus involves the State **(paragraph 1),** NGOs and political elites **(paragraph 2).**

Paragraph 1: Free distribution of mosquito nets by the State

As the "universal legatee of unresolved social problems"[240] , the State retains a monopoly in the process of stepping up the fight against malaria, which began in 2002. In fact, in 2003, the State was the first actor to carry out a campaign to distribute mosquito nets to the population of the Centre-Cameroon region. It is thus the central actor in the implementation of the policy of distributing mosquito nets to households. This centrality of the State was reflected in the campaigns to distribute LLINs to pregnant women and children under the age of five between 2003 and 2007 (A) and in the campaign to distribute LLINs on a mass scale to the population in 2016 **(B).**

A- Free distribution of Mil to pregnant women and children under the age of five

In the social health field in Centre-Cameroon, the State is the first actor to implement actions to fight malaria since the adoption of the *"Roll Back Malaria"* programme in 2002. At the time, it was the driving force behind the distribution of Mil to certain sections of the population, and was followed by several other players. In practical terms, the State's action in the field of malaria distribution in Centre-Cameroon is made possible by the "public action instruments"[241] at its disposal. By this we mean the "technical and social device that organises specific social relationships between public power and its addressees according to the representations and meanings it conveys"[242] .

The first action taken by the State to step up the fight against malaria in the Centre was the free distribution of Mil to households, the very first campaigns of which took place between 2003 and 2007. The main and only targets of these campaigns were pregnant women and children under five, considered to be the population groups most vulnerable to malaria, with 327 deaths recorded in 2000[243] . The

---

[240] See BIRNBAUM (P) and BADIE (B), Sociologie de 1 Etat, Pans, Grasset, 1979, p. 30.

[241] Christopher HOOD is the author of the seminal work on public policy instruments. In France, Frederic VARONE was the very first author to analyse public policy using instruments. See HOOD (C), Tools of Government. London, Macmillan, 1983. See also VARONE (F), Le Choix des instruments des politiques publiques. Bern, Haupt, 1998.

[242] Read HALPERN (C), LASCOUMES (P) and LE GALES (P) "L'mstrumentation et ses effets. Debats et mises en perspectives theoriques", in HALPERN (C), LASCOUMES (P) et LE GALES (P), (dir.), L'instrumentation de l'action publique. Op. cit... p.17.

[243] Figure compiled by the health information and planning department of the Centre regional public health delegation.

government thus mobilised "economic" and "incentive" instruments to implement its action, which was carried out in two sequences of public action. The first ran from 2003 to 2005 and covered only pregnant women (1), while the second ran from 2005 to 2007 and covered only children under the age of five **(2).**

1. Distribution of Mil to pregnant women (2003-2005)

The very first action taken by the State to step up the fight against malaria in the Centre region focused on "vector control", which consists of preventing the disease by distributing free Mil to households. To put this into practice, the government is using its incentive and communication tools, as it is seeking both the direct involvement of its citizens and an explanation of the measures taken. The first campaigns for the free distribution of Mil by the State took place in Central Cameroon between 2003 and 2005 and were aimed solely at pregnant women, one of the population groups most vulnerable to malaria, with twenty-four (24) cases of death in 2000[244] . In the region, malaria represents "a plausible potential danger" for this section of the population[245] , hence the circumscription of the Mil distribution perimeter around it between 2003 and 2005.

As a central player in the fight against malaria, the State laid the foundations for the process of stepping up the fight against the malaria problem by   organising the first "restricted"   campaigns for the distribution of LLINs to pregnant women in the geographical area of Centre-Cameroon. These first "restricted" campaigns took place between 2003 and 2005, and gave impetus to the free distribution of mosquito nets, which will continue until 2016. Specifically, in 2003, the government distributed 8,728 million; in 2004, 22,000 million; and in 2005, 44,014 million to pregnant women in the Centre region[246] . Throughout these different sequences, the state has held and "successfully claimed a monopoly on legitimate speech"[247] on the fight against malaria. The following table° 5 gives a clearer picture of the distribution process throughout the Centre region.

Table 5

Distribution of Mil to pregnant women from 2003 to 2005 in the Centre region

YEAR	Mil DISTRIBUTED	PREGNANT WOMEN	COVERAGE RATE
2003	8.728	129.782	6,7%
2004	22.000	133.286	16,5%

---

[244] Ibid.

[245] Plausible potential danger" occurs when people or living environments are affected by damage that is perfectly characterisable but whose precise nature and causes remain unknown. For more details on this concept see CALLON (M), LASCOUMES (P) et BARTHE (Y), <u>Agir dans un monde incertain: essai sur la democratic technique</u>. Paris, editions du Seuil, 2001, p. 39.

[246] Source: National Strategic Plan to Combat Malaria in Cameroon 2007-2010.

[247] A quotation borrowed from Fred EBOKO, who paraphrased Max WEBER to demonstrate that the hospital institution exerted a great influence on the fight against pathologies to the detriment of other players. Read EBOKO (F), "Introduction a la question du sida en Afrique. Politique publique et dynamiques sociales", Op. cit... PP. 14-15. In the context of this study, this expression refers to the fact that during the years 2003-2005, the State was the only actor involved in the field of malaria through the distribution of Milli.

| **2005** | 44.014 | 137.919 | 31,9% |

Source: National Strategic Plan to Combat Malaria in Cameroon 2007-2010

It is clear from this table that the rate of coverage of pregnant women with Mil is well below the target set by the strategic plan to combat malaria, which is 60%[248]. In 2003, only 8,728 Mil were distributed to 129,782 pregnant women, giving a coverage rate of 6.7%. In 2004, only 22,000 Mil were distributed to 133,286 pregnant women, giving a coverage rate of 16.5%. In 2005, only 44,014 Mil were distributed to 137,919 pregnant women, giving a coverage rate of 31.9%. Despite the revolution in the rate of mil coverage, it still falls far short of the targets set. The reasons for this are due to a number of socio-territorial constraints, which will be discussed in Chapter 3 of this report. In concrete terms, between 2003 and 2005, the process of distributing millet to pregnant women in the geographical area of Centre-Cameroon took place both in the city of Yaounde and in rural areas.

In concrete terms, since the adoption of the *"roll back malaria"* programme by the State of Cameroon, the very first action to step up the fight against malaria in the Centre's social health field has been the distribution of Mil to pregnant women in the city of Yaounde (Mfoundi). Yaounde was chosen because it is the heart of the region, in other words, the regional capital. The Mil distribution process in this geographical area was carried out over three successive years, namely 2003, 2004 and 2005, with pregnant women as the main target group. Specifically, in 2003, the government distributed 5,322 millet; in 2004, it distributed 12,034 millet and in 2005, 25,000 millet to pregnant women in the city of Yaounde. The distribution was carried out by health district according to the following tables° 6, 7 and 8.

Table 6

Number of Mil distributed to pregnant women by health district in Yaounde in 2003

HEALTH DISTRICTS	**NUMBER OF Mil DISTRIBUTED**
Biyem Assi	790
Green City	557
Djoungolo	1.610
Efoulan	640
Nkolbisson	715
Nkolndongo	1.010
TOTAL	**5.322**

Source: compiled by the author using data from the Centre regional public health delegation

The distribution of Mil to pregnant women in the six health districts of Yaounde was effective in 2003. The districts of Djoungolo, Nkolndongo and Biyem-Assi were the star performers, being "titanic" districts with a high concentration of pregnant women, unlike the districts of Cite Verte, Efoulan and Nkolbisson.

---

[248] The 2002-2006 national strategic plan for the fight against malaria sets the coverage rate for Mil at 60%.

Table 7

Number of Mil distributed to pregnant women by health district in Yaounde in 2004

HEALTH DISTRICTS	NUMBER OF MU DISTRIBUTED
Biyem Assi	2.334
Green City	985
Djoungolo	3.215
Efoulan	1.120
Nkolbisson	1.820
Nkolndongo	2.560
TOTAL	**12.034**

Source: compiled by the author using data from the Centre regional public health delegation

This table shows that millet was distributed to pregnant women in Yaounde in 2004. The number of millet distributed increased significantly compared with 2003. The districts of Biyem-Assi, Djoungolo and Nkolndongo once again came out on top, as these are districts with a high population concentration, and *a fortiori* a high concentration of pregnant women.

Table no. 8

Number of Mil distributed to pregnant women by health district in Yaounde in 2005

HEALTH DISTRICTS	NUMBER OF Mil DISTRIBUTED
Biyem Assi	3.615
Green City	2.242
Djoungolo	6.337
Efoulan	3.246
Nkolbisson	3.860
Nkolndongo	5.700
TOTAL	**25.000**

Source: compiled by the author using data from the Centre regional public health delegation

This table shows that millet was distributed to pregnant women in the various health districts of Yaounde in 2005. The number of millet distributed was once again significantly higher than in 2004. The districts of Djoungolo, Nkolndongo and Biyem-Assi once again obtained the largest number of Mil, given the high population density of pregnant women in these areas.

Table No. 9

Table summarising the distribution of Mil to pregnant women in Yaounde between 2003 and 2005

YEAR	Mil DISTRIBUTED	POPULATIONS PREGNANT WOMEN IN	COVERAGE RATE

		YAOUNDE	
2003	5.322	23.891	22%
2004	12.034	29.875	40,5%
2005	25.000	34.356	72,7%

Source: Compiled by the author using data from the National Strategic Plan to Combat Malaria in Cameroon 2007-2010.

The tables above show that the free distribution of Mil to pregnant women in the Yaounde districts was carried out in three distinct sequences: the first in 2003, the second in 2004 and the third in 2005. During these phases, the pace of distribution was essentially gradual, with the state distributing 5,322 mils in 2003, 12,034 mils in 2004 and 25,000 mils in 2005. Secondly, the rate of coverage of pregnant women with Mil in 2003 and 2004 fell short of the objectives set by the national strategic plan to combat malaria, which is 60%[249] . In 2005, the coverage rate was reached and even exceeded.

It should also be noted that the distribution of State Millet to pregnant women in the Centre region between 2003 and 2005 was not limited to the city of Yaounde, but extended to the surrounding area. Pregnant women in the twenty-four (24) districts of the region's Periphery have also rccи Mil de 1 Etat but with much less satisfactory results. Specifically, in this part of the Centre region, the government distributed 3,406 Millet in 2003, 9,966 in 2004 and 19,014 in 2005. The distribution was carried out by health district according to tables n° 10,11 and 12 below.

Table 10
Number of Mil distributed by the State in 2003 to pregnant women
in the 24 districts of the periphery

N°	Health districts	Number of Mil distributed
1	**Akonolinga**	290
2	**Awae**	103
3	**Ayos**	105
4	**Bafia**	257
5	**Ebebda**	55
6	**Elig-Mfomo**	98
7	**Eseka**	246
8	**Esse**	94
9	**Evodoula**	90
10	**Mbalmayo**	282
11	**Mbandjock**	100
12	**Mbankomo**	117
13	**Mfou**	204

---

[249]The 2002-2006 national strategic plan for the fight against malaria sets the coverage rate for Mil at 60%.

14	Monatele	167
15	Nanga-Eboko	301
16	Ndikimineki	50
17	Ngong-Mapubi	96
18	Ngoumou	209
19	Ntui	101
20	Obala	108
21	Okola	94
22	Sa'a	92
23	Soa	104
24	Yoko	43

Source: compiled by the author using data from the Centre regional public health delegation

This table shows that Mil was effectively distributed to pregnant women in rural districts in 2003. It can be seen that pregnant women in rural districts that are relatively desenclave received the most Mil. These were pregnant women living in districts either close to the city of Yaounde (Mbankomo, Awae, Mfou, Obala) or those corresponding to the departmental capitals (Akonolinga, Bafia, Eseka, Monatele, Nanga-Eboko, etc.). Pregnant women living in other, more isolated rural areas (Yoko, Ndikimineki, Esse, Ebebda, etc.) were not fully covered by the Mil system.

Table n°ll
Number of Millet distributed by the State in 2004 to pregnant women in the 24 outlying districts

N°	Health districts	Number of Mil distributed
1	Akonolinga	652
2	Awae	359
3	Ayos	397
4	Bafia	621
5	Ebebda	268
6	Elig-Mfomo	356
7	Eseka	508
8	Esse	277
9	Evodoula	334
10	Mbalmayo	657
11	Mbandjock	457
12	Mbankomo	419
13	Mfou	465
14	Monatele	477
15	Nanga-Eboko	511
16	Ndikimineki	257

17	Ngong-Mapubi	358
18	Ngoumou	415
19	Ntui	367
20	Obala	427
21	Okola	363
22	Sa'a	337
23	Soa	451
24	Yoko	233

Source: compiled by the author using data from the Centre regional public health delegation

As in 2003, pregnant women in rural health districts received гсси Mil in 2004. There was a significant increase in the number of millet distributed. Pregnant women living in rural districts close to the city of Yaounde or those corresponding to departmental capitals received the highest number of Millet. This is the case in the rural districts of Akonolinga, Bafia, Eseka, Nanga-Eboko, Obala, etc.

Table 12

Number of Millet distributed by the State in 2005 to pregnant women in the 24 outlying districts

N°	Health districts	Number of Mil distributed
1	Akonolinga	1470
2	Awae	632
3	Ayos	611
4	Bafia	1378
5	Ebebda	570
6	Elig-Mfomo	633
7	Eseka	1267
8	Esse	520
9	Evodoula	668
10	Mbalmayo	1524
11	Mbandjock	745
12	Mbankomo	788
13	Mfou	784
14	Monatele	736
15	Nanga-Eboko	825
16	Ndikimineki	504
17	Ngong-Mapubi	664
18	Ngoumou	784
19	Ntui	682
20	Obala	754
21	Okola	614
22	Sa'a	672

| 23 | **Soa** | 764 |
| 24 | **Yoko** | 425 |

As in 2003 and 2004, pregnant women in rural districts received State Millet in 2005, but the results were visibly better in terms of the number of Millet distributed. Pregnant women in districts close to the city of Yaounde or those corresponding to the departmental capitals obtained the highest number of Millet. This was the case for women living in the districts of Mbalmayo, Akonolinga, Bafia, Eseka, Nanga-Eboko, Bankomo and Obala.

The three tables above show that the State did indeed distribute Mil to pregnant women in the twenty-four outlying districts between 2003 and 2005. The results obtained can be seen in table 13 below.

Table 13
Table summarising the distribution of Mil to pregnant women in the outlying districts between 2003 and 2005

YEAR	Mil DISTRIBUTED	PREGNANT WOMEN IN OUTLYING DISTRICTS	COVERAGE RATE
2003	3.406	105.891	3,2%
2004	9.966	103.411	9,6%
2005	19.014	103.563	18,4%

This summary table shows that the results obtained from the distribution of Mil between 2003 and 2005 in the twenty-four peripheral districts are well below average, as we are a long way from the 60% coverage rate required by the national strategic plan to combat malaria in Cameroon.

If we try to make a comparison between the results obtained from the distribution of Millet to pregnant women in the city of Yaounde and those of the distribution to pregnant women in the outskirts, we can see that the sequence of distribution of Millet in the Centre region 20032005 was much more extended to the former at the expense of the latter. As the population of pregnant women in the periphery is higher than that of Yaounde, it would have been more rational and logical for the State to maximise the distribution of Millet in the twenty-four rural health districts. However, according to the tables above, the opposite result was obtained: in 2003, the distribution of Millet converted 22% of the population of pregnant women in Yaounde, compared with 3.2% of pregnant women living on the outskirts. In 2004, the distribution converted 40.5% of pregnant women in the city of Yaounde against 9.6% of pregnant women living in the outskirts. In 2005, the distribution converted

72.7% of pregnant women from Yaounde against 18.3% of pregnant women from the outskirts. These results contrast sharply with the population of pregnant women in the two areas. In 2003, the Yaounde districts had 23,891 pregnant women, in 2004 29,875 and in 2005 34,356. In the same years, the twenty-four rural districts had 105,891, 103,411 and 103,563 pregnant women respectively. The number of pregnant women in the peripheral health districts is much higher than that of pregnant women in the Yaounde districts. However, the Yaounde area obtained more Mil than the rural areas, so that in this area the coverage rate was even above average in 2005.

In short, the Millet distribution campaigns of 2003, 2004 and 2005 focused much more on pregnant women living in the city of Yaounde, to the detriment of those living in rural areas. Of the 8,728 millet distributed in 2003, 5,322 were distributed in the city of Yaounde. Of the 22,000 million distributed in 2004, 12,034 million were distributed in Yaounde. And of the 44,014 millet distributed in 2005, 25,000 millet were distributed in Yaounde[250] . This means that more than half of the millet was distributed in the health districts of Yaounde and the rest on the outskirts! A community health worker (CHW) who was involved in these campaigns to distribute MU to pregnant women explains:

"The mosquito nets distributed to pregnant women in the Centre region in 2003, 2004 and 2005 only concerned the health districts in and around Yaounde, such as Obala, Mfou, Awae and Mbankomo. This was because the number of agents responsible for distribution was very low, and the outlying districts were very isolated at the time"[251] .

This goes some way to explaining why, of the 129,782 Mil planned for distribution to pregnant women in 2003, only 8,728 were actually distributed, with a low coverage rate of 6.7% for all thirty health districts!

The phase of distributing Mil to pregnant women between 2003 and 2005 was a failure, as the coverage rates (6.5%, 16.5% and 31.9%) were well below the 60% expected and desired by the Cameroonian authorities. The empirical reasons for this failure will be presented and analysed in chapter three of this thesis. The second phase of net distribution by the state ran from 2005 to 2007 and targeted children under five in the region.

2. Distribution of Mil to children under five (2005-2007)

The second phase of the intensification of the fight against the malaria problem in the Centre region, based on "anti-vector" control, was again carried out by the State. This took place between 2005 and 2007. During this phase, Mil was no longer distributed to pregnant women, but to children under the age of five, considered to be the most vulnerable section of the population, with 303 cases of death recorded in 2000[252] . This sequence of actions to combat malaria is monopolised by the State, which holds and "successfully claims a monopoly on

---

[250] Figures provided by the Centre regional public health delegation.
[251] Interview with a community health worker living in Yaounde.
[252] Figure compiled by the health information and planning department of the Centre regional public health delegation.

legitimate speech"[253] over the process of distributing Mil to children under five.

In concrete terms, the distribution of Millet to children under the age of five was spread over three successive years, namely 2005, 2006 and 2007. During these years, 511,335 children under the age of five were registered in the region, i.e. 155,667 in all the Yaounde districts and 355,668 in the outlying districts. Only 351 Mil were made available to children in these two areas, with a virtually negative coverage rate. This can be seen in table° 14 below.

Table 14

Results of the free distribution of Mil to

children under five in 2005, 2006 and January 2007 in the

Centre region

Mosquito nets distributed in 2005, 2006 and January 2007	351 Mosquito nets
Population of children under five	511,335 children under the age of five
Coverage rate for children under five	0,1%
Deviation from 2002 targets	-59,9%

Source: National Strategic Plan for Malaria Control in Cameroon 2007-.
2010

This table shows that the coverage rate for children under five in Mil is practically zero, at 0.1%, which is 59.9% below the 2002 target of 60%. The distribution process covered children in both Yaounde and the outlying districts.

Specifically, in 2005, 2006 and 2007, the State distributed 281 Millet to children under the age of five in Yaounde's health districts. Table 15 below shows the results of this Mil distribution campaign.

Table 15

The distribution of Mil to children under the age of five in the

Yaounde districts between 2005 and January 2007

Health districts	**Population of children under five**	**Number of Mil distributed in 2005, 2006 and January 2007**	**Coverage rate**
Biyem Assi	33.721	56	0,2%
Green City	9.360	16	0,2%
Djoungolo	38.805	77	0,2%
Efoulan	16.456	23	0,1%
Nkolbisson	18.847	40	0,2%

---

[253] A quotation borrowed from Fred EBOKO, who paraphrased Max WEBER to demonstrate that the hospital institution exerted a great influence on the fight against pathologies to the detriment of other players. Read EBOKO (F), "Introduction a la question du sida en Afrique. Politique publique et dynamiques sociales", Op. cit... pp. 14-15. In the context of this work, this expression refers to the fact that during the period 2003-2005, the State was the only actor involved in the field of malaria through the distribution of Mil.
[271] The 2002-2006 national strategic plan for the fight against malaria sets the coverage rate for Mil at 60%.

| Nkolndongo | 38.478 | 69 | 0,2% |
| TOTAL | **155.667** | **281** | **0,2%** |

The table shows that 281 nets were distributed to children under five in the six health districts of Yaounde in 2005, 2006 and 2007. It also appears that the coverage rate for children under five is practically zero, at 0.2%, i.e. well below the targets set by the 2002-2006 national malaria control strategic plan[271] . Out of a total of 155,667 children under the age of five counted in 2005, 2006 and 2007, only 281 Mil were made available, representing a coverage rate of 0.2%.

In addition, during 2005, 2006 and 2007, the State also distributed free Mil to children under the age of five in outlying areas, i.e. children who do not live in Yaounde. Seventy (70) Mil were made available to 355,668 children under the age of five. The results of this free millet distribution campaign can be seen in table° 16 below.

Table no. 16

Distribution of Mil to children in the twenty-four districts of the Periphery between 2005 and January 2007

Health districts	Population of children under five	Number of Mil distributed in 2005, 2006 and 2007	Coverage rate
Akonolinga	27.101	10	0,03%
Awae	8.675	00	00
Ayos	11.120	00	00
Bafia	23.567	10	0,04%
Ebebda	7.603	00	00
Elig-Mfomo	7.436	00	00
Eseka	24.953	10	0,04%
Esse	8.627	00	00
Evodoula	8.460	00	00
Mbalmayo	28.764	10	0,03%
Mbandjock	24.123	05	0,02%
Mbankomo	12.820	10	0,08%
Mfou	11.040	05	0,04%
Monatele	14.500	00	00
Nanga-Eboko	26.623	00	00
Ndikimineki	10.333	00	00
Ngong-Mapubi	9.355	00	00
Ngoumou	23.172	00	00
Ntui	14.610	00	00

Obala	12.400	05	0,04%
Okola	9.424	00	00
Sa'a	11.850	00	00
Soa	8.270	05	0,06%
Yoko	10.842	00	00
TOTAL	**355.668**	**70**	**0,01%**

Source: compiled by the author using data from the National Strategic Plan for Malaria Control in Cameroon 2007-2010.

This table shows that a number of outlying health districts did not receive any Millet during the action sequence for distributing Millet to children under five from 2005 to 2007. It also appears that out of 355,668 children under the age of five in outlying areas, only 70 millet were made available, with a coverage rate of 0.01%. It was also noted that only the rural health districts corresponding to the departmental capitals or those located close to the city of Yaounde were able to see some of their children under the age of five receive a Mil. This was the case in the health districts of Akonolinga, Bafia, Eseka, Mbalmayo, Mbandjock, Mbankomo, Mfou, Obala and Soa. This shows that the State's action has been very minimal in the peripheral areas and, from this point of view, rather than being reflexive, the State has been "minimal" because it has not managed to deploy itself in the peripheral areas of the Centre region.

In short, it is clear that the objectives set for the distribution of Mil to children under five between 2005 and January 2007 were not achieved, because out of 511,335 children under five in the whole of the Centre region, only 351 Mil were made available, a coverage rate of 0.1%! The main reason for this fiasco can be found in the words of the regional public health delegate for the Centre region:

"The government had planned to make mosquito nets available to every child under the age of five, in line with the 2002 targets. But from 2005 onwards, the nets were produced in insufficient quantities and the possibility of achieving the targets was practically nil. The policy now is to distribute one mosquito net per household with at least one child under the age of five.

It is clear from these comments that the reason for the very low rate of Mil coverage for children under five in the 2005, 2006 and January 2007 distribution campaigns is the glaring inadequacy of the Mil. Consequently, the distribution policy was to distribute a single Mil to households with at least one child under the age of five, when we know that there are households with up to 10 children in this age group[254][255]. This is why, out of 511,335 children under the age of five, only 351 Mil were made available to them! It should also be pointed out that the distribution campaign in question was limited to the health districts in and around Yaounde, which is why the outlying areas had a Mil coverage rate of 0.01%.

In conclusion, the first steps to step up the fight against malaria were taken by the

---

[254] Interview with Charlotte Moussi, regional public health delegate for the Centre region, on 13 January 2021.
[255] For more details on this subject, read the Plan Strategique National de Lutte contre le Paludisme au Cameroun 2007-2010. Op. cit... pp. 58-59.

State between 2003 and 2007. To make this "undertaking" operational, the State used its "economic" and "incentive" instruments, distributing Mil free of charge to certain social strata of the population and encouraging them to acquire and use them through communication. During the period 2003-2007, the free distribution of millet took place in two clearly identifiable sequences of public action. The first sequence ran from 2003 to 2005 and was aimed solely at pregnant women. With twenty-four (24) deaths recorded in the year 2000, the State saw fit to devote the first phase of Mil distribution to this social stratum. This distribution campaign covered both the urban districts of Yaounde and the outlying rural districts. In 2003, the State distributed 8,728 millet, in 2004 22,000 millet and in 2005 44,014 millet, with coverage rates of 6.7%, 16.5% and 31.9% respectively. During this first phase of millet distribution, the state was much more active in urban than in rural areas. This explains why the coverage rates were well below the targets set by the 2002-2007 national strategic plan to combat malaria in Cameroon. The reasons for this failure are partly due to social constraints which have considerably limited the action of the state agents in charge of distributing Mils to pregnant women in the geographical area of Central Cameroon. These social constraints will be analysed in Chapter 3 of this thesis.

The second phase of Mil distribution by the State took place just after the first and lasted through 2005, 2006 and January 2007. This phase of distribution was aimed solely at children under the age of five, who, alongside pregnant women, are the most vulnerable segment of the population, with 303 cases of death recorded in 2000. To avoid what Fred EBOKO describes as an "epidemiological shock"[256] , the government has taken the initiative of carrying out a campaign to distribute Mil free of charge to this section of the population. In principle, this distribution campaign should have extended to both urban and rural areas, but it was confined to the health districts of Yaounde and those of the capitals of the departments of the slaves or those located near Yaounde. The main reason for this state of affairs is that the State was faced with a crying shortage of Mil during this public action period, and the strategy consisted in distributing the little money that was available. Mil to households with at least one child under the age of five in selected health districts. That's why, out of 511,335 children under the age of five counted in the 2005, 2006 and 2007 corns, only 351 Mil were distributed, for a coverage rate of 0.1%. Which is an abject failure! The result will not be the same when it comes to the "mass" campaign to distribute LLINs to households in 2016.

B- Mass distribution of LLINs to households

From April 2016, a campaign will begin for the mass distribution of free LLINs to households previously sensitised by individual and collective actors as part of the behaviour change communication as described above. During this process, the

---

[256] In other words, the exponential increase in deaths due to the spread of an epidemic or disease. See EBOKO (F), *"Introduction a la question du sida en Afrique. Politiques publiques et dynamiques sociales"*, in KEROUEDAN (D) and EBOKO (F), Politiques publiques du sida en Afrique. Travaux et documents n°61-62, CEAN, IEP Bordeaux, 1999.

State remains the dominant player and mobilises new material resources: Long-Acting Impregnated Mosquito Nets (LLINs). Unlike the Mil, which was distributed to certain sections of the population between 2003 and 2007, the LLIN is unique in that it is a mosquito net treated with an insecticide (deltamethrin) directly incorporated into the polyester fibre at the time of manufacture. It has a lifespan of around three (03) years, is effective for up to twenty (20) washes and does not need to be re-impregnated (washing is enough to reactivate it). The LLIN is therefore a more reliable, more effective and longer-lasting tool for protection against mosquito bites than the LLIN. In this respect, the mass free distribution of LLINs to all households in the Centre region by the government from April 2016 onwards is a decisive step towards stepping up the fight against malaria. This operation represents the implementation of strategic priority 2 of the PSNLP, devoted to vector control. It is now indisputable that the female anopheles is the vector par excellence of malaria transmission through the plasmodium it inoculates into the bodies of healthy subjects. Vector control is becoming the surest way of avoiding any risk of malaria through the mass distribution of LLINs to households.

As a key player in the fight against malaria, the French government is once again taking action to combat the vector-borne disease during the April-May 2016 period, in accordance with the information memorandum issued by the French Ministry of Health.

MINSANTE of 15 April 2016[257] . To operationalise this new sequence of public action, the State is mobilising human and material resources to make its "undertaking" plausible. An enumeration team and community health workers (CHWs) have been mobilised to carry out the distribution of LLINs in the social field. A total of 2,311,626 LLINs were distributed to households throughout the Centre region between April and May 2016 by the government[258] , with more or less satisfactory results in terms of household coverage rates. In practical terms, the process of distributing LLINs by the State was carried out in two stages. The first took place from Monday 18 to Saturday 30 April 2016 and was devoted exclusively to "enumeration surveys" **(1)**, followed by the second from Monday 16 to Tuesday 31 May 2016, devoted to the actual distribution of the LLINs **(2)**.

1. The household enumeration phase (18-30 April 2016)

The prerequisite for achieving "universal coverage of the population with LLINs" was the setting up of so-called "enumeration" surveys, whereby MINSANTE agents were sent out into the field to enumerate the households that were to receive one or more LLINs from the State. In other words, the "mass" distribution of LLINs to households was preceded by a general door-to-door enumeration/awareness-raising operation by an enumeration team[259] . In the social health field of Centre-Cameroon, this operation took place between 18-30 April

---

[257] The full text of this information memorandum can be found in appendix 4 to this report.
[258] Figures provided by the Centre regional public health delegation.
[259] See the document entitled Cameroon: enquete post campagne sur l'utilisation des moustiquares impregnees d'insecticide a longue duree d'action 2016/2017. Op. cit... P. 72.

2016, with MINSANTE agents carrying out an exhaustive household enumeration prior to any distribution of LLINs. The enumeration process was carried out door-to-door, with coupons/tickets being handed out to households, followed by a brief awareness-raising session. Receipt of a coupon/ticket by a household was the *sine qua non* condition for receiving an LLIN during distribution. As a result, households that had not received a coupon/ticket could not claim an LLIN at the time of distribution. Everything was organised in such a way as to ensure that the LLIN distribution campaign was as rational as possible. The counting itself consisted of counting the number of people living permanently in each household. In the Mfoundi department, i.e. the city of Yaounde, almost 2,600,000 people were counted[260] . Vouchers were given to households after counting the number of members living there and the number of sleeping spaces. The logic was then that of one coupon for two members of the same household.

Overall, around three quarters of households (76.9%) were enumerated at the end of the process[261] . In the city of Yaounde, 68.5% of households were enumerated, compared with 81.7% on the outskirts[262] . One study showed that out of a sample of 1,006 households in the city of Yaounde, 628 had received a coupon between 1830 April 2016[281] . The same study showed that out of a sample of 491 households in the centre of Yaounde, without Yaounde, 385 had received a coupon during the same period[282] . This suggests that not all households in the Centre region have received coupons/tickets, which are the "key" to accessing the LLIN.

According to the final report published by MINSANTE, PNLP and INS in December 2017 on the 2015-2016 LLIN campaign, 32.7% of households in Yaounde and 18.3% of those in the outlying areas of the Centre did not receive a coupon/ticket in April 2016[283] . This has acted as a brake on the process of "mass" distribution of LLINs, as we shall see in the next paragraph. The least that can be said at present is that there are several reasons why households in the Centre's social health field are not receiving coupons. These reasons, as well as the percentage of households that have not received coupons, can be seen in the following table° 17.

Table 17

**Reasons for the non-receipt of coupons/tickets by households
during the 18-30 April 2016 enumeration campaign in
the Centre region**

Reasons given	Percentage of households	Percentage of households

---

	not receiving a coupon in Yaounde	not receiving a coupon in the Centre region (excluding Yaounde)
The team ran out of coupons	3,9%	0,8%
We weren't at home at the time	76,4%	60,2%
We have refused	2,1%	0,0%
Other reasons	12,1%	24,2%

Source : compiled by 1 author from data taken from Rapport final, MINSANTE, PNLP, INSdedecembre 2017, Op. cit... p. 73.

This table shows that there were several reasons why some households did not receive coupons during the enumeration process on 18-30 April 2016. Firstly, on arrival at some households, the enumeration team ran out of coupons. 3.9% of households in Yaounde and 0.8% of those in the Centre excluding Yaounde did not receive coupons for this reason. Secondly, when the enumeration team arrived in some homes, the members were absent for various reasons. 76.4% of households in Yaounde and 60.2% of those in the outskirts did not collect coupons for this reason. Some households in Yaounde (2.1%) systematically refused to take a coupon. Other reasons linked to social constraints prevented some households from receiving coupons. Elies are higher in Peripherie than in Yaounde. We shall present them in the next chapter of this thesis. The least that can be said at present is that the household enumeration phase was not a complete success in view of the fact that some households did not have гссн coupons/tickets. An empirical study carried out in December 2017[284] in the geographical area of Centre-Cameroon showed that out of a sample of 1,006 households randomly selected in the city of Yaounde, only 628 said they had received a ticket between 18-30 April 2016 and 329 said they had not received one for the reasons given in the table above. It follows that 49 households did not feel concerned by the enumeration campaign and *a fortiori* by the free distribution of LLINs because they had not been previously sensitised by the CCC. For this reason, they did not receive any coupons. This explains why, in Yaounde as a whole, the percentage of households counted between 18 and 30 April was 68.5%, which seems low if we want to achieve "universal coverage of LLINs", the percentage of which is set at 80% by the NMCP. It should nevertheless be noted that in the Centre region as a whole (Yaounde+periphery), 81.7% of households were enumerated during the April 2016 campaign, despite the cases of non-receipt recorded. The enumeration phase was followed by the distribution of LLINs between 16 and 31 May 2016, which marked a real step-up in the fight against malaria.

2. The actual "mass" distribution of LLINs (16-31 May 2016)

Two weeks after the end of the household enumeration phase, the government began the "mass" distribution of LLINs, a real phase in the intensification of the campaign. The aim of the "mass" distribution was to improve the availability and use of LLINs. In accordance with the instructions on the coupon, it was planned to distribute one LLIN for every two people in each household, rounded up to the next unit, depending on the size of the household[285] . This was the objective to be achieved if the public authorities wanted to ensure "universal coverage of the population with LLINs". To achieve this, "distribution points" for LLINs were set up by MINSANTE agents in areas open to the public. Each household, armed with one or more coupons, was expected to go to the nearest distribution point to y collect their LLIN(s). Upon presentation of a coupon, each household was entitled to receive an LLIN for two people - one for each person.

[284] Read the document entitled Cameroon: enquete post campagne sur l'utilisation des moustiquares impregnees d'insecticide a longue duree d'action 2016/2017. Op. cit... p. 73

[285] Ibid. p. 77.

per household. In total, 2,311,626 LLINs were actually distributed to households in the Centre-Cameroon geographical area out of the 3,584,478 LLINs expected, for a universal coverage rate of 64.4%, well below the target set by the NMCP, which should have been 80% by the end of the operation[263] . The reasons for this failure will be analysed below. The least that can be said at present is that, compared with previous campaigns for the distribution of mosquito nets by the State (2003-2005 and 2005-2007), the rate of coverage with LLINs achieved in 2016 in the region can be considered a real success, even though the final objective was not achieved. In concrete terms, in the centre of Yaounde (Mfoundi), 1,432,511 LLINs were made available to households, compared with 879,115 LLINs for the 24 outlying districts. The distribution of LLINs by health district can be seen from tables° 18 and 19 below.

Table no. 18

Number of LLINs distributed in Yaounde districts by the State in 2016

HEALTH DISTRICTS	NUMBER OF MILDA DISTRIBUTED
Biyem Assi	222.438
Green City	87.380
Djoungolo	440.256
Efoulan	90.000
Nkolbisson	162.223
Nkolndongo	430.214
TOTAL	**1,432,511 MILDA**

Source: compiled by the author using data from the Centre regional public health delegation

This table shows that a number of LLINs were made available to the populations of Yaounde's health districts by the State during the free mass distribution of LLINs in

[263] Figures obtained from the Delegation Regional de la sante publique du Centre.

2016. A total of 1,432,511 LLINs were provided to households in the six districts of Yaounde. The districts of Djoungolo, Nkolndongo and Biyem-Assi received the highest number of LLINs, simply because they are "titanic" health districts with high population concentrations.

Table 19
Number of LLINs distributed in the twenty-four
peripheral health districts
by the State in 2016

N°	Health districts	Number of LLINs distributed
1	Akonolinga	52.344
2	Awae	37.451
3	Ayos	36.278
4	Bafia	50.112
5	Ebebda	25.510
6	Elig-Mfomo	35.144
7	Eseka	47.338
8	Esse	25.231
9	Evodoula	25.444
10	Mbalmayo	52.123
11	Mbandjock	37.623
12	Mbankomo	42.027
13	Mfou	59.368
14	Monatele	44.569
15	Nanga-Eboko	49.604
16	Ndikimineki	24.112
17	Ngong-Mapubi	14.544
18	Ngoumou	39.000
19	Ntui	29.567
20	Obala	30.543
21	Okola	35.623
22	Sa'a	35.132
23	Soa	35.500
24	Yoko	14.928
	TOTAL	**879.115**

Source: compiled by the author using data from the Centre regional public health delegation

These two tables show that between 16-31 May 2016, 1,432,511 LLINs were distributed to households in the city of Yaounde, compared with 879,115 to those in the surrounding area. The department of Mfoundi alone accounted for almost half of all the LLINs distributed in the Central region, which is understandable given that the districts of Yaounde are "titanic districts" with almost 400,000 inhabitants in the districts of Ndjoungolo, Nkolndongo and Biyem-Assi, unlike the "Lilliputian districts" on the outskirts, some of which have barely 20,000 inhabitants, such as the district of Yoko. In total, 2,311,626 LLINs were distributed throughout the Centre region at the end of the operation, out of the 3,584,478 LLINs actually expected. This means that not everyone who should have received an LLIN was able to get one, as the "universal coverage" target (80%) was not achieved.

The reason given by most households for not collecting their LLINs was that they had lost their coupons. "*On the day I had to go to the distribution point to collect my LLINs, I couldn't find my coupon, I didn't know where I'd put it*"[264] confided a father living in the Biyem-Assi neighbourhood. The coupon was so small and light that it was difficult for households to keep it. What's more, there was no "duplicate" that could be used as proof if the original coupon was lost. This was one of the main limitations of the mass distribution of LLINs by the government in 2016. Some households also mentioned the problem of the distance between their homes and the distribution point and the time taken at the distribution point to receive an LLIN as reasons for not receiving it. A survey showed that in outlying areas, people had to spend an average of 430 FCFA and in the city of Yaounde 218 FCFA to get to the distribution point[265] . As a result, 41.6% of households in Yaounde and 23.5% of households in the outlying areas did not receive any State-issued LLINs in 2016[266] . Table 20 below summarises the results of the mass distribution of LLINs in the social health sector in Centre-Cameroon between 16 and 31 May 2016.

Table 20
Results of the 2016 mass distribution of LLINs
in the Centre region

	Yaounde	Centre (excluding Yaounde)
Percentage of households having received at least one LLIN	56,8%	75,6%
Percentage of households that have not received an LLIN	41,6%	23,5%
Average number of LLINs per household	1,5	2,4
Percentage of households	41,2%	47,4%

---

[264] Interview with Etienne Nyemb, father of a family living in the Biyem-Assi district on 13 August 2020.
[265] Read the document entitled <u>Cameroon: enquete post campagne sur l'utilisation des moustiquares impregnees d'insecticide a longue duree d'action 2016/2017</u>. Op. cit... p. 77.
[266] Ibid.

having received at least one LLIN for every two people usually living in the same household		
Percentage of households receiving fewer LLINs than expected	20,3%	24,5%
Percentage of households receiving the planned number of LLINs	16,5%	18,1%

Source: compiled by 1 author using data from the Final report, MINSANTE, PNLP, INSdedecembre 2017,Op. cit... p. 73.

From this table, it appears that the "mass" distribution campaign of LLINs by the State in 2016 was not a success with regard to the percentage of households that did not have гсси LLINs (41.6% in Yaounde and 23.5% in the peripheral zone), in terms of the percentage of households having received at least one LLIN (56.8% in Yaounde and 75.6% in the outlying areas) and in terms of the percentage of households having received at least one LLIN for every two people (41.2% in Yaounde and 47.4% in the outlying areas). It also appears that the rate of coverage with LLINs was higher in the periphery than in Yaounde, as 41.6% of households in Yaounde did not receive an LLIN compared with 23.5% in the periphery. Most members of Yaounde households said they were at work at the time of distribution. All in all, the universal coverage rate for LLINs in the Centre region was 64.4%, well below the target set by the NMCP and MINSANTE.

In short, in terms of its deployment in the social field, the State appears to be the central actor in the process of distributing mosquito nets in the geographical area of Centre-Cameroon. It became involved in this process in 2003, just after the adoption of the *Roll Back Malaria* programme. The first campaigns to distribute mosquito nets to households took place over the period 2003-2007 and were aimed primarily at those sections of the population most vulnerable to malaria, i.e. pregnant women and children under the age of five. These initial campaigns were generally unsuccessful. To rectify the situation, the government is implementing a mass distribution campaign of LLINs to households between 16 April and 31 May 2016, with generally satisfactory results in terms of population coverage. While the central role of the State in the net distribution process is clear, it should be noted that, in line with the "multisectoral" dynamic, other players are involved in the process.

Paragraph 2: Free distribution of mosquito nets by NGOs and political elites

Although the State is a central player in the process of distributing free ITNs/MILDAs to households, it does not monopolise the process. Other actors are involved in the process of stepping up the modern fight against the malaria problem through the free distribution of ITNs/MILDAs. These actors are NGOs **(A)** and political elites **(B).**

A- Free distribution of mosquito nets by NGOs

With the aim of stepping up the modern fight against malaria in the Centre-Cameroon region, several local NGOs entered the vector control process between 2007 and 2011. Alongside the State, they became major players in the process of distributing mosquito nets. In other words, during this period of public action against malaria, these collective actors[267] followed the path constructed and marked out by the state by distributing free mosquito nets to certain groups of households. In this way, they complement state action by co-producing public action to combat the problem of malaria. To carry out their 'enterprise' in the social field, local NGOs mobilise and mobilise material resources. The most visible of these are the "Cameroon Association for Social Marketing" (1) and the "Alliance for Malaria Prevention" **(2).**

1. Free distribution of mosquito nets to vulnerable people in the city of Yaounde by ACMS

The Association Camerounaise pour le Marketing Social (ACMS) is an NGO founded in 1996[268] , with its head office in Yaounde and a regional branch in Mballa II-Dragage. ACMS is a major player in the fight against malaria in Central Cameroon, and more specifically in Yaounde. Elie has the material resources and knowledge to carry out its work, which involves distributing "super moustiquaire" insecticide-impregnated mosquito nets and "Bloc" impregnation kits.

Between 2007 and 2010, ACMS distributed free mosquito nets to pregnant women, children under the age of five and disadvantaged children in the city of Yaounde. On 10 July 2007, ACMS launched its first campaign to distribute its "super mosquito nets" free of charge to pregnant women in Yaounde district hospitals. At the end of this campaign on 30 July 2011, 1,500 super mosquito nets were distributed to 1,500 pregnant women in the hospitals of Yaounde's six health districts[269] . According to Mr ETONO Pierre, a member of the NGO's Yaounde regional branch, *"the aim of this campaign was purely philanthropic. The aim was to protect pregnant women in hospital from mosquito bites"[270]* .

The second free distribution campaign of "super mosquito nets" and "Bloc" impregnation kits by ACMS was extended to households with at least one child under the age of five in the city of Yaounde. This campaign was a continuation of the action taken by the State, which had already made 281 Mil available to children under the age of five in Yaounde between 2005 and 2007. *"This campaign*

---

[267] Collective actors refer to legal entities and are still referred to as corporate actors because they bring together individuals who share a common point of interest. For Patrick HASSENTEUFEL, collective actors can be administrative entities, government bodies, political organisations, interest groups, communities of experts, etc. See HASSENTEUFEL (P), Sociologie politique : l'action publique. Op. Cit... p. 172.

[268] Following a collaboration agreement between the Social Marketing Programme in Cameroon (PMSC) and Population Services International (PSI), from which it receives technical and financial assistance. See www.infosdafrique.com.

[269] Figures provided by the Yaounde regional branch of ACMS.

[270] Interview by ACMS in Yaounde.

complemented the government's January 2007 campaign"[271] confirms Mr ETONO Pierre. Between 15 January 2008 and 10 February 2008, ACMS distributed "400 super mosquito nets to 400 households with at least one child under the age of five"[272] .

Finally, the third "super mosquito net" distribution campaign by ACMS was aimed solely at disadvantaged children in the city of Yaounde. On 27 August 2010, ACMS decided to distribute 2,000 long-lasting impregnated mosquito nets (LLINs) free of charge to 24 care facilities for disadvantaged children in Yaounde and the surrounding area[273] . On the same day, members of ACMS launched the *"United Against Malaria"* week at the "Foyer de I'esperance de Yaounde"[274] , providing this social rehabilitation centre with almost 200 LLINs[275] . By the end of the campaign week on 02 September 2010, exactly 2,000 LLINs had been distributed to 24 social reintegration centres in Yaounde and the surrounding area[276] .

Between 2007 and 2010, ACMS continued the fight against malaria initiated by the government in 2003 by distributing "super mosquito nets" and "Bloc" impregnation kits. This sequence of action against malaria was extended only to vulnerable people, i.e. pregnant women, children under the age of five and disadvantaged children. Through this series of actions, ACMS is playing a major role alongside the government in stepping up the fight against malaria. Another NGO will be working in the same direction from 2012.

2. Mass" distribution of LLINs by 1'APP

The Alliance for Malaria Prevention (APP) is an NGO that brings together more than 40 partners, including government institutions, private companies, private sector organisations and faith-based and humanitarian organisations. Each of these partners makes a valuable contribution to the partnership by helping countries to achieve the objectives of the *Roll Back Malaria* (RBM) programme through increasing the number of people owning and using an LLIN. In the Centre-Cameroon region, this NGO will be taking part in the fight against malaria from Monday 02 January 2012 with a mass distribution campaign of LLINs, known as the *"Hang-Up"* campaign. According to the head of the family health division at ACMS, which was a partner in this campaign, *"the Alliance for Malaria Prevention's hang-up campaign was a complement to the mass distribution of LLINs by the State. This partnership group intervened in Cameroon because it realised that the public authorities had not achieved universal coverage of the population with LLINs"[277]* . The *"Hang-Up"* campaign follows on from the

---

[271] Ibid.

[296] Read *Cameroon Tribune* of Monday 30 August 2010.

[297] This is a rehabilitation centre for street and prison children located in the Dakar district at[295] Ibid.

Yaounde and is currently run by Father Alfonso RUIZ.

[275] Read *Cameroon Tribune* of Monday 30 August 2010.

[276] Ibid. [295] Ibid.

[277] Interview with Mr Jean-Christian YOUMBA, Head of the Family Health Division at ACMS on 07 February 2022.

government's mass distribution of LLINs, and strengthens the fight against malaria in the Centre region. In practical terms, APP's intervention in the field of malaria is divided into two phases: the distribution phase and the LLIN hanging-up phase.

Between 02 and 14 January 2012, almost 2,281,583 LLINs were distributed in Central Cameroon by the APP as part of the *"Hang-Up"* campaign[278] . This campaign represents another step towards stepping up the fight against malaria. The process of distributing LLINs was a mixed one, with the APP setting up both distribution points and door-to-door campaigns. In the city of Yaounde, the distribution of LLINs to households was carried out via distribution points after an enumeration operation. Each member of a household who went to a net distribution point had to first present a coupon showing the number of LLINs he or she was to receive. In total, 1,832,144 LLINs were distributed to households in Yaounde at the end of the process on 14 January 2012.

In outlying areas, the APP distributed LLINs to households using the door-to-door strategy. Community health workers (CHWs) were mobilised in the various health areas to ensure the distribution of mosquito nets as part of the "home-based malaria management" policy (PECADOM). Between 02 and 14 January 2012, households in the outlying health districts received LLINs from CHWs with stocks. The provision of LLINs to households took place after a brief interview lasting around fifteen minutes. During this interview, the CHWs were asked to identify the people at risk of malaria, the number of people per household, the number of sleeping spaces per household and any cases of malaria. This is what the PECADOM implemented by l'APP in the outlying areas of Centre-Cameroon between 02-14 January 2012 was all about. The distribution process followed the logic of one LLIN for every two people living in the same household, or alternatively one LLIN per sleeping space. Once the nets had been distributed to a household, a chalk mark was put on the wall of the house, indicating that the members of that household had already been visited by the CHWs. This procedure was used in the Ntui and Yoko health districts, where a chalk cross could be seen on the concessions that had received LLINs from the APP.

In total, in the social health field of Centre-Cameroon, the *"Hang-Up"* LLIN campaign made it possible to distribute almost 2,281,583 LLINs in the various health districts, with a coverage rate of 85%[279] (in other words, universal coverage was achieved). According to Mr Jean-Christian YOUMBA, *"the success of the LLIN 'hang-up' campaign implemented by the alliance for the prevention of malaria must be sought in the mixed strategy it has adopted, i.e. the opening of distribution points in health centres and door-to-door distribution"*[303] . The mobilisation of community health workers for door-to-door distribution of LLINs has been particularly beneficial for households in rural areas. According to a local man in Yoko, *"a chalk mark was clearly visible in front of every house after 14*

---

[278] Read the <u>Cameroon mission report on the LLIN hang-up campaign</u> on 27 May 2012.
[279] Ibid.

January 2012". In other words, all the households in the town of Yoko have had CHWs deliver mosquito nets to their homes!

As the majority of households have LLINs, the APP wanted to encourage their use. For this reason, it organised an "LLIN Hang-Up" campaign in all the health districts of the Centre region. In fact, two weeks after the *"Hang-Up"* campaign, the APP organised an "LLIN hanging" campaign between 1[er] and 15 February 2012 throughout the Centre region, with the emphasis on outlying areas. The hanging campaign was organised in response to the fact that households were not using, or were misusing, their LLINs. In fact, several CHWs in the field found that some households were unable to hang up the LLINs they had received from the government. What is the point of organising a mass distribution campaign for LLINs if people don't know how to use them? In order to rationalise the fight against vectors, the APP thought it advisable to carry out an LLIN hanging campaign in which health workers would go round households to educate the masses.

Between 1[er] and 15 February 2012, 68 CHWs per health district went round households to hang up the LLINs. The demonstration was carried out in households whose members had difficulty hanging a mosquito net. Equipped with clones and hammers, the CHWs appointed by the PPA first talked to the heads of each household before carrying out the hanging demonstration. The purpose of the interview was to -

[303] Interview with Mr Jean-Christian YOUMBA, Head of the Family Health Division at ACMS on 07 February 2022.

check whether each member of the household sleeps under a mosquito net or, failing that, whether all sleeping areas are covered by a mosquito net. If the members of a household expressed any difficulty in hanging their LLINs, the CHWs would do so on the spot using the hanging tools at their disposal. In this way, *"in the Biyem-Assi health district, 106 households benefited from the expertise of the CHWs to be able to see the LLINs hung in their rooms between 1[er] and 15 February 2012"[280]* .

The innovative hanging campaign is breathing new life into vector control by "forcing" people to sleep under a mosquito net. Before the campaign was launched, many households in Central Cameroon did not sleep under an LLIN, even though they had one. Households cited the lack of time to hang up the nets and the difficulty of carrying out the operation. With the *"Hang-Up"* campaign, the APP is putting an end to the "non-use" of LLINs. This is bound to give new impetus to the process of stepping up the fight against malaria. The political elites were also involved in the process of distributing mosquito nets to households.

B- Free distribution of mosquito nets by political elites

Political elites are political actors, i.e. actors who are directly involved in electoral

[280] Interview with a community health worker from the Biyem-Assi health district who worked for l'APP between 2012 and 2017 on 13 February 2021. The interviewee preferred to remain anonymous.

competition to occupy positions of power at national and local level[281] . These are elected representatives who are heavily involved in shaping local public action and who live 'for' politics, or 'from' politics[282] . Elections represent a resource that is both specific and decisive for this category of actors in shaping public action, and it is also a strong determinant of the strategies pursued by these actors in the context of public policy[283] .

The 2007-2011 period was also characterised by the intrusion of political elites into the field of malaria in the Centre region. Together with NGOs, they played a part in stepping up the modern fight against the disease, starting with the implementation of the 'multi-sectoral' principle. Benefiting from the mandate of the local population, the political elites of the Centre-Cameroon geographical area, by virtue of their proximity to the population, encourage the local direction of malaria control policies. They play a key role in the territorialisation of malaria policies. They have a perfect grasp of people's needs and difficulties, and are capable of adjusting malaria control policy to the realities of their area. When we talk about political elites, we are referring to political players such as communal executives and members of parliament. Their actions in the field of vector control are only visible during election periods, which is why they are described as "strategic". Local elected representatives use two types of resources to have a significant impact on the area in which they are active: political resources and time resources.

Between 2007 and 2011, the political elites of Central Cameroon took part in the fight against malaria, following the path set out by the State. In this sequence, their action is essentially based on the distribution of Millet to those in need. To make this action plausible on the ground, they use the political and time resources at their disposal. In this way, they exploit their position as politicians (as elected representatives) and the electoral "moment" to have a greater impact on the social field. On the eve of elections, in this case the parliamentary elections of 2007 and the presidential elections of 2011, the political elites of Centre-Cameroon donated Mil to certain health facilities. These political elites are communal executives **(1)** and members of parliament **(2)**.

1. The donation of mosquito nets to health facilities by local authorities

Communal authorities in the Centre region have been interfering in the process of distributing millet since 2007, on the occasion of Selection municipale. They use time and political resources to implement this policy of distributing millet to the local population. Their action is therefore essentially strategic in that they calculate the electoral moment to deploy on the ground. Their position as local elected representatives also predisposes them to work more closely with the local population and to have a considerable impact on them. In the commune of Yaounde

---

[281] Read HASSENTEUFEL (P), Sociologie politique : l'action publique. Op. cit... p. 245.
[282] Read WEBER (M), Le savant et le politique. Pion, 1959, p. 111.
[283] See HASSENTEUFEL (P), Ibidem, p. 245.
[308] Interview with Mr ANABA Philippe, former employee of Yaounde 6 Town Hall on 22 April 2020.

sixieme, for example, Mr Robert Atangana distributed Mil to the hospitals in the Biyem-Assi health district on 18, 19, 20 and 21 July 2007. Malaria is becoming an electoral campaign issue on which politicians are trying to win over voters.

Mr Robert Atangana distributed two hundred (200) mosquito nets to twenty (20) health centres in the Biyem-Assi health district, as reported by a former employee of the Yaounde Sixth Town Hall: *"On the eve of the municipal elections, the outgoing mayor, Robert Atangana, made two hundred mosquito nets available to twenty health centres in the Biyem-Assi district to protect the sick in these centres from mosquito bites. The operation was welcomed by the directors of the health facilities"*[308] . Twenty health facilities were supplied with mosquito nets at the end of the distribution operation on 21 July 2007, as shown in table 21 below.

Table 21

Donation of Mil to health facilities in the Yaounde 6 commune by the CPDM Mayor,

Mr Robert ATANGANA

Distribution date	Health training	Number of Mil distributed
18/07/2007	Presidential Guard infirmary	10
18/07/2007	HD of Biyem-Assi	10
18/07/2007	Biyem-Assi Clinic	10
18/07/2007	Complexe medical sota	10
19/07/2007	Alma Ata healthcare practice	10
19/07/2007	Camp Yepyap infirmary	10
19/07/2007	Well-being health centre	10
19/07/2007	Centre de sante la difference	10
19/07/2007	Cabinet de soins Marc Vivien FOE	10
20/07/2007	Cabinet de soins saint joseph	10
20/07/2007	Centre de sante la charite	10
20/07/2007	Centre de sante confiance plus	10
20/07/2007	Cabinet de soins mere Theresa	10
21/07/2007	Centre de sante jesus care	10
21/07/2007	Cabinet de soins delivrance	10
27/07/2007	Sainte Marie Elisabeth nursing home	10
21/07/2007	Accacia care practice	10
21/07/2007	Cabinet de soins st Michel	10

21/07/2007	C.S performance	10
21/07/2007	Cabinet de soins le reconfort plus	10
TOTAL	**20**	**200**

Source: compiled by the author from data supplied by Mr ANABA Philippe, former employee of Yaounde City Council 6.

The mosquito nets were distributed by a group of ten young people recruited and paid by the outgoing mayor. This social action by Mr ATANGANA Robert, far from being an act of humanism, was essentially strategic because it was part of his electoral campaign and that of the CPDM. For him, it was a question of seducing the medical staff as well as the population as a whole, who saw in him a magnanimous man who would hold the reins of Yaounde Sixth Commune for one more term. We can therefore easily state that it was on the basis of his involvement in the fight against vectors that Mr ATANGANA Robert was re-elected as CPDM Mayor of Yaounde Sixth Municipality at the end of the Municipal Selection of 22 July 2007.

In addition, another political player made a name for himself in the commune of Yaounde premier on the occasion of Selection municipale by donating Mil to health facilities. This was Mr Emile ANDZE-ANDZE, who made the fight against malaria one of his campaign themes for the 22 July 2007 election. His position as a local elected official and outgoing mayor has enabled him to be more active in the social arena. When distributing mosquito nets to health facilities, he mobilised political and time resources. Specifically, Mr Emile ANDZE-ANDZE distributed three hundred mosquito nets in fifteen health centres in the Djoungolo district on 17, 18 and 19 July 2007. The operation was made possible by a group of nine young people deployed to the main health centres in the Yaounde I district. The young people, who were not students, lived in the Etoudi, Mballa II and Ndjoungolo neighbourhoods. Distribution was carried out by health facility according to the following table° 22.

Table 22
Donation of Mil to health facilities in Yaounde Ier by
CPDM mayor
Emile ANDZE ANDZE

Date of donation	Health training	Number of Mil distributed
17/07/2007	EPC Djoungolo Hospital	30
17/07/2007	Jamot Hospital	40
17/07/2007	HD D'Olembe	30
18/07/2007	Nyom 2 victory	10
18/07/2007	Elig-Dzoa humanitarian medical centre	15

18/07/2007	Nkoleton Catholic dispensary	20
18/07/2007	Etoudi CMMR	20
18/07/2007	C.S notre dame d'Emana	20
19/07/2007	Localite Mballa II	20
19/07/2007	Bastos Clinic	30
19/07/2007	Adventist dispensary	20
19/07/2007	Emana CSI	15
19/07/2007	Social Health Centre	10
19/07/2007	CS la rosee	10
19/07/2007	CS perseverance	10
Total	**15**	**300**

Source: compiled by the author on the basis of data provided by Mr ELEMVA'A Bertrand, who was one of the young students involved in the process of distributing Mil launched by the Mayor, Mr ANDZE.

The distribution was carried out in the form of donations by nine young people recruited for this purpose. A box of plastic-coated mosquito nets was given to the directors or heads of the target health facilities. Mr Bertrand ELEMVA'A described the distribution process in the following terms:

"There were nine of us in total, in three groups of three people each. The nets were donated on 18, 19 and 20 July 2007. Every morning around eight o'clock, we went to the town hall to pick up boxes of mosquito nets, and then we went to the target health facilities to donate them. We carried badges that read 'Yaounde I commune, donation from Mr Emile Andze'"[309] .

Mr ANDZE-ANDZE's donation to the health centres was a purely electoral operation, given the timing of its launch. This political player was on the campaign trail, and knew full well that malaria is a major concern for the people of the 1 arrondissement of Yaounde 1[er] and that the health facilities there are not very well equipped with Mil. Not being able to satisfy all the health facilities in the district, the most prominent ones were targeted by the CPDM politician for donation. Far from being a humanist or humanitarian action, Mr ANDZE-ANDZE's donation of MUs was essentially "strategic". His re-election as head of the municipal executive of the district of Yaounde 1[er] was partly the result of this strategic act.

In the rural areas of the region, several politicians also focused their election campaigns on the fight against malaria, both during the 2007 municipal elections and the 2011 presidential elections. In the rural commune of Ntui, the CPDM, led by Mr Jerome ONDOBO MONO, proceeded to distribute MUs in the middle of the Ntui market to the traders who were there and to the district hospital on 18 July 2007, on the eve of the municipal elections. The pretext for this distribution was the high number of malaria cases in the Ntui health district. As the area is forested, people are constantly exposed to mosquito bites. While this was the obvious motive, it should be noted that the distribution of the Mil to the co-incinerators and health workers of Ntui also had a latent motive, that of renewing the CPDM at the head of the municipal executive of the Ntui district. In reality, this was a "strategic" action in view of the "moment" chosen for the distribution of the Mil. This strategic approach continued in 2011, during the presidential election campaign. In September 2011, mosquito nets wrapped in plastic with the CPDM logo were distributed to households in the rural communes of Ntui, Ngambe-Tikar and Yoko. The distribution was carried out by a group of young people from the Mbam-et-Kim OJRDPC in homes located on the roadside. Recipients were reminded each time that it was a gift from the CPDM. The operation was piloted by the municipal executives of the above-mentioned communes. From this point on, we can already see the close link between elections and the distribution or donation of mosquito nets to the population.

2. Donation of mosquito nets by members of parliament

From 2007 onwards, as elections approached, a number of political players used malaria to build their electoral campaigns in order to win the votes of the population. This political phenomenon is much more noticeable during local elections such as the legislative and municipal elections. In the department of Mfoundi, as in the city of Yaounde, a veritable "system of action" is formed[284] in the process of implementing public action to combat malaria, with the entry on the scene of political elites who are de facto or unofficial implementers. Yves MENY and Jean-Claude THOENIG write that :

"implementation structures a field of action in which many players are involved, both those to whom the task of execution is formally devolved - for example, external State services - and more distant players, in principle not concerned, but who are mobilised because implementation represents a challenge for them or for their representatives"[285].

The political elites of the Centre region in general, and of the Mfoundi department in particular, belong to the category of 'distant' actors, because they are sporadically involved in the field of malaria.

On the eve of the parliamentary elections on 22 July 2007, the Biyem Assi and Cite-Verte district hospitals received a large donation of malaria control kits from the various parliamentary candidates. The kits include mosquito nets, insecticides, medicines and IPT for pregnant women. The words of the general supervisor of the Biyem-Assi district hospital are enlightening: *"In 2007, this hospital received donations from the parliamentary candidates. These included boxes of medicines, insecticides and mosquito nets"*[286] . Over and above its humanitarian nature, the action of the Mfoundi elites is essentially strategic and even utilitarian, since the ultimate objective is to be re-elected, hence the mobilisation of time resources that enable them to act only during the electoral period. The MPs Vincent de Paul EMAH ETOUNDI and Paul-Eric DJOMGOUE were partly re-elected as MPs on

---

[284] A system of action corresponds to the structure of implementation as devised by the German sociologist Ranate MAYNTZ. It corresponds to a set of relationships established between players at the time of implementation in the field.
[309] Interview on 22 April 2022 with Bertrand Elemva'a, who was one of the young students involved in the distribution of mosquito nets launched by Mr Andze on 18, 19 and 20 July 2007.

[285] See MENY (Y) and THOENIG (J-C), Politiques publiques. Op. cit... pp. 253-254.
[286]Interview with the general supervisor of Biyem-assi district hospital on 3 February 2021.

the basis of the use of this resource. Similarly, at the start of the electoral campaign for the 2011 presidential election, a number of parliamentarians from the Centre region used malaria-related themes to win over the electorate. The most emblematic and charismatic figure in this exercise was Marie-Rose NGUINI EFFA, MP for Mefou-et-Akono. Her position as president of the "network of parliamentarians, populations and development" and as a member of the "association of dynamic women of Cameroon" gives her "authority" in the fight against epidemics in pregnant women and children under the age of five. She distributed Mils to this social stratum in the town of Ngoumou on 6 October 2011, three days before the presidential election. The timing chosen by this political player was strategic: the aim was to seduce voters into voting for the CPDM.

In rural areas, the issue of combating malaria was also put on the table during the 2011 presidential election campaign. The mosquito net then became an instrument of seduction for voters, because through it, the parties were able to arouse the sympathy of the electorate. On 07 October 2011, the Honourable Pierre MGBATOU, considered to be the "spokesman" for the young people of Mbam-et-Kim, donated 700 mosquito nets, which were distributed to households in the town of Ntui. The people of this geographical area saw this act as a "gesture" of magnanimity on the part of the CPDM. But the operation was far more strategic than humanitarian. First of all, the Honourable Pierre MGBATOU, who enjoyed a strong social and political position, was strategically the one person who could carry out such an action better than anyone else. Secondly, it was necessary to wait until the electoral period, and moreover the electoral campaign period, to make donations of mosquito nets to the population. The problem of malaria has been co-opted by the political elites for electoral purposes.

In short, NGOs and political elites participate in their own way in the process of distributing mosquito nets to households. In a way, they complement the action of the State, which occupies a central position in this process. NGOs such as ACMS and APP are stepping up the fight against malaria through their "therapeutic activism". Political elites are also involved in the process of institutionalising the fight against malaria through their various actions in the run-up to an election. Thus, from 2007 onwards, communal executives and members of parliament became involved in the fight against malaria through the distribution of mosquito nets to the population.

In conclusion, from 2002 onwards, the trajectory of the fight against malaria in the geographical area of Centre-Cameroon underwent a sharp bifurcation due to both endogenous and exogenous changes in this sub-system. On the endogenous level, it is the involvement of new players in the field of malaria that is driving the dynamics of change. Exogenously, it was the adoption by the United Nations of the Millennium Development Goals (MDGs)[287] and the approval by the Cameroonian

---

[287] A great deal of attention is devoted to the fight against malaria in Africa.

authorities of the "*Roll Back Malaria*" programme[288] that gave new impetus to the fight against malaria. In concrete terms, from 2002 onwards, the fight against malaria became tougher, and now obeys a multi-actor dynamic, given the adoption of the principle of "multisectoriality". The traditional methods used prior to the implementation of this principle are giving way to modem methods. The result was a major institutional change. This change is in fact driven by a wide range of actors, over a wide range of asynchronous timeframes, and by a wide range of social processes, also asynchronous, affecting different dimensions or aspects of the institutions or institutionalisation processes[289] . The fight against malaria is undergoing an *institutional conversion,* i.e. the new players involved in the field of malaria since 2002 are reorienting the fight against this disease towards new mandates, new objectives, new challenges and new functions. The fight against malaria is now based on new administrative methods, practices and techniques. These are being put into practice in the social field by stepping up communication to change people's behaviour, increasing the power of the public authorities and distributing mosquito nets free of charge. A new *design* and new cognitive scripts are thus associated with the fight against malaria.

The actors driving change and the intensification of the fight against malaria in the geographical area of Central Cameroon are both individuals and groups. They are mobilising public action instruments and resources to give new meaning to the fight against malaria and to take significant action on the social front. Thus, the vast household awareness campaign as part of the CCC, public decision-making and the free distribution of LLINs are the new malaria control "policies" implemented in the Centre region and encouraged by the WHO and the PSNLP. Throughout these processes, the state retains a monopoly and is the central and dominant player.

<u>CONCLUSION OF PART ONE</u>

The fight against the malaria problem in the Centre-Cameroon region was institutionalised through a sequence of actions whose trajectory underwent bifurcation points between 1919 and 2020. Two main phases can be identified in the process of institutionalising the fight against malaria: 1919-2002 and 2002-2020. Between 1919 and 2002, the fight against malaria in the geographical area of Centre-Cameroon was marked by the privileged use of traditional methods. During the colonial period, i.e. from 1919 to 1960, the fight against malaria was characterised by the privileged intervention of non-specialist actors and by the strong presence of traditional and prophetic practices. French military doctors had the privilege of intervening in the field of malaria between 1919 and 1950, and

---

[288] This is the "Roll Back Malaria" programme, which aims to eradicate malaria in the short term. It was set up in 1998 at the instigation of Dr Gro Harlem Brundland, and is a consortium comprising the World Bank, UNDP, UNICEF and WHO. It was approved in Cameroon in 2000 and its committee began operating effectively on 29 July 2002 following decision N°0334/MPS/CAB. For more details see MOULIOM MOUNGBAKOU (LB), Op. cit... pp. 137-157.

[289] See DEMAILLY (L), GIULIANI (F) and MAROY (C), "Le changement institutionnel: processus et acteurs", p. 3.

implemented what was known as "itinerant offensive" medicine, both in the field of prevention and treatment. Physical force was used to force local populations to receive modem health care. But with the advent of the emancipation movements around the 1950s and the exponential growth of the local population, French military medicine was put on the back burner, in favour of traditional and prophetic medicine. At this point, the fight against malaria underwent its first fork in the road since 1919. Local populations were now using local therapies and traditional rites to combat malaria. For the indigenous people, decolonising Cameroon meant decolonising its health system. Traditional and prophetic practices replaced military medicine, which was run by non-specialists, until 1960, when the fight against malaria was slightly modernised.

Between 1960 and 2002, the fight against malaria was slightly modernised as a result of the arrival of health professionals and the relative rationalisation of traditional medicine. The professionalisation of the fight against malaria is reflected in the "managerial reforms" of the health system in Central Cameroon. These reforms led to the creation of reference hospitals, a large-scale medical school and health districts. Hospitals and doctors became the major players in the institutional system for combating malaria, and the lay people who had previously been involved found themselves sidelined. The result is a situation of institutional change that Kathleen THELEN illustrates using the concept of 'institutional layering'[290].

Moreover, as part of this process of slight modernisation in the fight against malaria, traditional medicine, once considered an archaic method, resurfaced and became relatively rationalised around 1990. With the ratification of the Alma Ata Declaration by the State of Cameroon, and the issuing of provisional certificates for the practice of traditional medicine to certain herbalists, traditional medicine acquired a "legal" character. This mechanism of reproduction and readaptation of previous institutions in the policies of the present corresponds to what Paul PIERSON calls 'path *dependence*'. From this point of view, the processes of institutional development in the fight against malaria that began in 1960 in the Centre-Cameroon region are subject to the constraints imposed by the regions chosen earlier. Hence the "institutional inertia" that has perpetuated traditional medicine despite the dynamics of institutional change and modernisation in the fight against malaria. Consequently, rather than being complete, the modernisation of the fight against malaria that began between 1960 and 2002 is relative, slight and limited. We had to wait until 2002 to see a real change in the dynamics of the institutionalization of modern malaria control. In fact, with the adoption of the "*Roll Back* Malaria" programme in 2002, the modern fight against the disease was stepped up and tightened, putting a definitive end to traditional methods of combating malaria.

---

[290]See THELEN (K), "Comment les institutions evoluent perspectives d'analyse comparative historique", Op. cit... p. 30.

In fact, from 2002 onwards, the trajectory of the fight against malaria in the geographical area of Centre-Cameroon underwent a sharp bifurcation due to both endogenous and exogenous changes in this sub-system. On an endogenous level, it is the involvement of new players in the field of malaria that is driving the dynamics of change. Exogenously, it was the adoption by the United Nations of the Millennium Development Goals (MDGs)[291] and the approval by the Cameroonian authorities of the *"Roll Back Malaria"* programme[292] that gave new impetus to the fight against malaria. In concrete terms, from 2002 onwards, the modern fight against malaria became tougher, and henceforth obeyed a multi-actor dynamic in view of the adoption of the principle of "multisectoriality". The traditional methods that had been used before the implementation of this principle gave way to modem methods. The result is a major institutional change. This change can be brought about by a wide range of actors, at different and asynchronous times, and by different social processes, also asynchronous, affecting different dimensions or aspects of the institutions or institutionalisation processes in different ways[293] . The fight against malaria is undergoing an *institutional conversion,* i.e. the new players involved in the field of malaria since 2002 are redirecting the fight against this disease towards new mandates, new objectives, new challenges and new functions. The fight against malaria is now based on new administrative methods, practices and techniques. These are being put into practice in the social field by stepping up communication to change people's behaviour, increasing the power of the public authorities and distributing mosquito nets free of charge. A new *design* and new cognitive scripts are thus associated with the fight against malaria.

The actors driving change and the intensification of the fight against malaria in the geographical area of Central Cameroon are both individuals and groups. They are mobilising public action instruments and resources to give new meaning to the fight against malaria and to take significant action on the social front. Thus, the vast household awareness campaign as part of the CCC, public decision-making and the free distribution of LLINs are the new malaria control "policies" implemented in the Centre region and encouraged by the WHO and the PSNLP. Throughout these processes, the state retains a monopoly and is the central and dominant player.

It should be pointed out, however, that the process of institutionalising the fight against malaria, which began in 1919 during the colonial period and was reinforced in 2002, has undergone a number of setbacks. In fact, several constraints punctuate this process in the Centre region, thus hindering the equilibrium of the dynamics of

---

[291] A great deal of attention is devoted to the fight against malaria in Africa.

[292] This is the "Roll Back Malaria" programme, which aims to eradicate malaria in the short term. It was set up in 1998 at the instigation of Dr Gro Harlem Brundland, and is a consortium comprising the World Bank, UNDP, UNICEF and WHO. It was approved in Cameroon in 2000 and its committee began operating effectively on 29 July 2002 following decision N°0334/MPS/CAB. For more details see MOULIOM MOUNGBAKOU (LB), Op. cit... pp. 137-157.

[293] See DEMAILLY (L), GIULIANI (F) and MAROY (C), "Le changement institutionnel: processus et acteurs", Op. cit... p. 3.

change and leading to periods of major disruption. As a result, the process of institutionalising modern malaria control has had ambivalent effects in this geographical area.

Part II

A dynamic of institutionalisation marked by a "punctuated equilibrium" and leading to ambivalent effects

The dynamics of the institutionalization of malaria control in the social and health field in Central Cameroon are subject to numerous tribulations. A long period of stable equilibrium in the establishment of the model malaria control system, as described in the preceding chapters, is followed by short periods of radical change, which punctuate the equilibrium of the dynamic of change that began with the colonial period in 1919. Frank BAUMGARTNER and Briand D. JONES[320] illustrate this reality by using the concept of '*punctuated* equilibrium'[321]. According to the initial intuition of these two authors, public policy is historically characterised by long periods of stability punctuated by short periods of radical change[322]. From a heuristic point of view, the punctuated equilibrium model aims to explain why public policy tends to experience long periods of stability, punctuated by abrupt periods of radical change. These radical changes constitute constraints on the dynamics of the institutionalisation of a public policy.

In the Centre-Cameroon region, although the dynamics of the institutionalization of malaria control have been stable for a long time, it should be pointed out that this gradual and linear dynamic of change is also subject to periods of abrupt change which put a brake on the equilibrium sought in this process. In other words, the equilibrium of the dynamics of institutionalisation of the model fight against malaria is constantly punctuated when policy images and policy locations undergo radical changes. In the geographical area of the Centre, it is the different games and interactions between institutional actors and the culture of the populations that

---

[320] Read BAUMGARTNER (ER) and JONES (B.D), <u>Agendas and Instability in American Politics</u>. Chicago, IL: The University of Chicago Press.

[321] Baumgartner and Jones found the name of their model, punctuated equilibrium, in a theory of paleontology that rejected gradualism, the theory according to which species evolve slowly through successive slight mutations. Like their paleontological colleagues, Baumgartner and Jones were opposed to a linear vision of the revolution of systems. They thus developed an explanatory model that calls into question the fact that policies evolve only in a regular and gradual way. Read MASSE (J M), *"Introduction au modele de I'equilibre ponctue: un modele pour comprendre la stabilite et les changements radicaux en politiques publiques"*, Montreal, Quebec: National Collaborating Centre for Healthy Public Policy, 2018, p. 2.

[322] Ibid., p. 1.

constantly punctuate the equilibrium of the institutionalization of the model malaria control, thus constituting strong tribulations to this dynamic **(chapter 1).** Consequently, the process of institutionalising the fight against malaria, which began in 1919, produced ambivalent effects on the social field **(Chapter 2).**

Chapter 3

The tribulations of the dynamics of institutionalisation of the modern fight against malaria

The process of institutionalising the fight against malaria in the Centre-Cameroon region is fraught with tribulations. According to Daniel KUBLER and Jacques DE MAILLARD[323], any public policy, when it is implemented, encounters 'backlash' or difficulties that prevent it from becoming institutionalised in the social arena. They write that "a policy is never implemented in a vacuum; it enters a context characterised by problems, structures and constraints that predate it, which provokes reactions in return. These reactions in turn influence the activities of the implementers. Implementation is thus seen as a dynamic process marked by a coming and going"[324]. Thus, according to the two political science professors, when public policy is being implemented, it is subject to 'constraints', which refer to anything that can provoke 'feedback' and which significantly bias or influence the actions of the implementers, making it difficult to translate a policy into concrete action. It is from this point that the process of institutionalising a public policy undergoes tribulations of various kinds.

In the context of this work, the term "tribulations" refers to "constraints", i.e. any difficulty or vicissitude of an institutional and social nature that punctuates the equilibrium of the dynamics of the institutionalization of the modern fight against malaria in the geographical area of Centre-Cameroon. To better analyse these tribulations, we will mobilise two methodological approaches to neo-institutionalism, namely rational choice neo-institutionalism and sociological neo-institutionalism. Because rational choice neo-institutionalism is based on the idea that public action, like human action, is explained by the strategies of individuals seeking to maximise their personal interest[325], it will enable us to demonstrate how the strategic interactions between health institutions and the actors who incite them constitute strong tribulations to the dynamics of institutionalisation of the model

---

[323] See KUBLER (D) and DE MAILLARD (J), Op. cit... p. 73.

[324] Ibid. p. 73.

[325] Read HALL (P.A) and TAYLOR ROSEMARY (C.R), "Political science and the three neo-institutionalisms", Op. cit... pp. 476-481.

fight against malaria. Sociological neo-institutionalism[326] will enable us to demonstrate that the 'social conventions', 'cognitive scripts' and 'cultural codes' shared by the members of the geographical space of Centre-Cameroon are all tribulations to the dynamics of the implementation of the model fight against the malaria problem.

The following lines will be devoted to an analysis of the tribulations that punctuate the dynamics of the institutionalization of modern malaria control in the Centre-Cameroon region. We will thus highlight the institutional tribulations leading to 'institutional complexity' on the one hand **(Section 1)** and the socio-cultural and territorial tribulations leading to 'territorial and social complexity' on the other **(Section 2).**

SECTION I

The tribulations Пёез to health institutions: ' institutional complexity'

Although Durkheim's influence was, as we know, decisive in the legal doctrine of the 3^e Republic, the rediscovery by sociologists of the concept of institution seems relatively recent. For some, institutions remain "established social forms", for others "processes by which society organises itself", this second approach having a dynamic dimension which is lacking in the first[327] . In the context of this work, the first meaning will undeniably be the most credible for a better understanding of the concept of health institutions, because institutions are inherently stable and static because of their tendency to endure. With Emile DURKHEIM, we define institutions "as social phenomena, impersonal and collective, presenting permanence, continuity, stability"[328] . From this, we can define a health institution as any impersonal and collective organisation or structure that is characterised by permanence, continuity and stability, designed to satisfy a public health need. When a society has a large number of 'busy' institutions, the stage is set for 'complexity'.

Yannis PAPADOPOULOS[329] defines the term complexity as "the presence of a large number of elements with a high degree of interdependence, making it both necessary and difficult to achieve a high degree of cooperation between them"[330] . The professor of political science develops three levels of complexity, including institutional complexity, which is characterised by the presence of complex decision-making structures[331] .

The definition of "institutional complexity" used in this study goes beyond that given by Professor Yannis PAPADOPOULOS. For us, institutional complexity

---

[326] Ibid. pp. 481-486.
[327] See QUERMONNE (J-L), "Les politiques institutionnelles. Essai d'interpretation et de typologie", in GRAWITZ (M) and LECA (J) (eds.), <u>Les politiques publiques</u>. Op. cit... p. 62.
[328] Quoted by QUERMONNE (J-L), Idem. p. 63.
[329] See PAPADOPOULOS (Y), <u>Complexite sociale et politiques publiques</u>. Op. cit... p. 48.
[330] Ibid. p. 48.
[331] See PAPADOPOULOS (Y), Idem. p. 58.

refers to the existence of a multiplicity of health institutions (and therefore a multiplicity of players) which are so intertwined that coordination between them seems impossible. The 'sociometric' study[332] of the social health field in Centre-Cameroon shows that there is a multitude of health institutions and training courses which, instead of cooperating with each other for joint action, tend to consider themselves as autonomous entities without taking account of externalities. This leads us to analyse institutional complexity using a neo-institutionalist approach described as 'actor-centred institutionalism'[333] . In his celebrated book *Games real actors play. Actor-centered institutionalism and policy research,* Fritz SCHARPF gives the quintessence of this approach in these terms: "*the approach proceeds from the assumption that social phenomena are to be explained as the outcome of interactions among intentional actors - individual, collective, or corporate actors, that is - but that these interactions are structured, and the outcomes shaped, by the characteristics of the institutional settings within they occur*"[334] .

Thus, 'actor-centred institutionalism'[335] is an unquestionable approach for anyone wishing to analyse 'institutional complexity' on the basis of interactions between actors with strategic behaviours when these interactions are structured by the institutional context in which they take place. More specifically, with regard to the conduct of public policy, actor-centred institutionalism focuses on the constraints and opportunities that structure the room for manoeuvre of the actors involved, and thus shapes the formulation of public policy[336] . Like the neo-institutionalism of rational choice, this approach privileges those aspects of human behaviour that are instrumental and oriented towards a strategic calculation[337] . This is why it is necessary to analyse the tribulations of the dynamics of the institutionalization of malaria control linked to health institutions. Elie is close to the strategic analysis developed by Michel CROZIER and Erhard FRIEDBERG[338] and will be logically used in this section because it tends to consider that there is no social action, that there is no collective structure without freedom of the actors and, consequently, without power relations[339] . These power relations between institutions and players in the social and health field in the Centre-Cameroon region skew the way in which public decisions are taken and implemented by central public players. In this

---

[332] A study that looks at a social field through the prism of an organisation that only exists through the interplay of its players.

[333] This is a conceptual framework for the analysis of public policy which has been developed since the 1970s by a group of German researchers around Ranate Mayntz and Fritz Scharpf, both specialists in public administration. This current considers that social phenomena are the result of interactions between actors with strategic behaviours, but that these interactions are structured by the institutional context in which they take place and which, consequently, also determines the result of these interactions. Based on this postulate, this conceptual framework can be likened to rational choice institutionalism.

[334] Read SCHARPF (F), Games real actors play. Actor-centered institutionalism and policy research, Boulder (CO), Westview, 1997, p. 1.

[335] Corresponding to the neo-institutionalism of rational choice.

[336] Read KUBLER (D) and DE MAILLARD (J), Analyser les politiques publiques. Op. cit... p. 119.

[337] Read HALL (P.A) and TAYLOR ROSEMARY (C.R), "La science politique et les trois neo-institutionnalismes", Op. cit... p. 472.

[338] Le CROZIER (M) and FRIEDBERG (E), L'acteur et le systeme. Op. cit... p. 433.

[339] Ibid. p. 433.

respect, they constitute major tribulations to the dynamics of institutionalisation of the modern fight against malaria in the social health field of Centre-Cameroon. They are reflected in the permanent entanglement of malaria treatment facilities **(Paragraph 1)** and in the lack of cooperation between health facilities **(Paragraph 2).**

Paragraph 1: The intertwining of malaria control structures

By entanglement, we mean the existence within an organisation of several players who influence several decision-making poles to the point where the implementation of decisions is biased. Entanglement situations arise within an organisation when the political-administrative arrangement (PAA) is lacking on the part of the players. It is therefore the failure of a political-administrative arrangement that leads to situations of entanglement within an organisation, since an APA "covers not only public actors, but also private actors who can be assimilated to them by virtue of the fact that they are invested with public powers and who, by virtue of this delegation of responsibility, participate on an equal footing in the production of concrete actions (outputs) linked to the public policy in question"[340] . ABS therefore produces positive coordination conducive to the implementation of concerted, joint and coherent public action.

To speak of the intertwining of structures involved in the fight against malaria in the Centre region is still to view this social health field through the prism of an organisation made up of a multitude of health institutions which operate without any regard for cooperation and coordination. This geographical health area is made up of a regional public health delegation, thirty health districts and 1,303 health facilities (regional and similar hospitals, district hospitals, district medical centres, integrated health centres). Schematically, its pyramid structure is shown in diagram° 1 below.

Schema n° 1

The pyramid structure of the social health field in Central Cameroon

---

[340] Read KNOEPFEL (P), LARRUE (C), VARONE (F) and SAVARD (J-F), <u>Analyse et pilotage des politiques publiques</u>. Op. cit... p. 241.

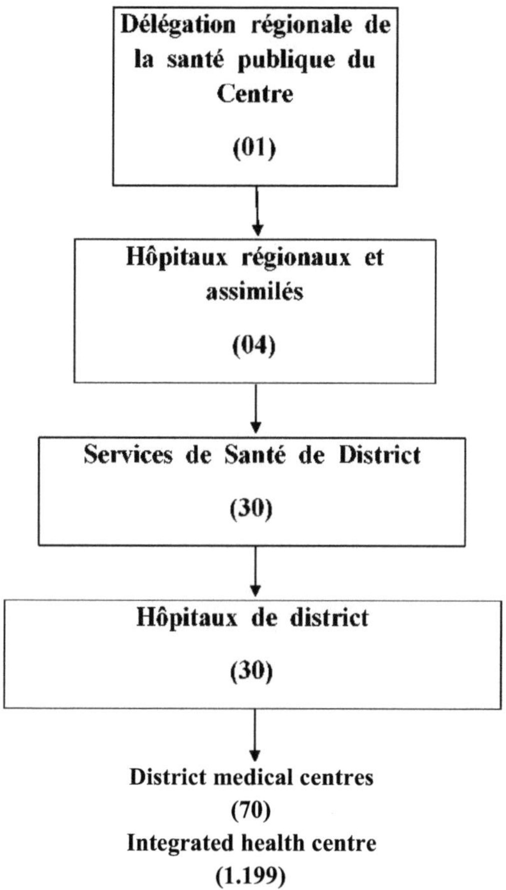

Source: compiled by the author from data provided by the Delegation Regionale de la Sante Publique du Centre.

Seen through the prism of an organisation, the structuring of the social health field in Central Cameroon reveals that it is a 'multi-actor' organisation that must obey a high degree of coordination to avoid situations of political entanglement. This structuring, because it highlights several health structures, makes it possible for power relations to develop and provides the basis for their permanence[341] . Hence the need for the 'multi-actors' involved in the implementation process to take account of a 'multi-actor' political-administrative arrangement to ensure that the actions produced are not contradictory. Indeed, "the greater the number of actors, the more ABS should have precise mechanisms for negotiating and establishing the competences and procedures required for the coordinated management of multiple activities"[342] . In the absence of such an ABS in an organisation, it will be nothing

---

[341] Read CROZIER (M) and FRIEDBERG (E), L'acteur et le systeme. Op. cit... p. 78.
[342] See KNOEPFEL (P), LARRUE (C), VARONE (F) and SAVARD (J-F), Op. cit... p. 245.

more than a world of conflict, and its functioning the result of clashes between the contingent, multiple and divergent rationalities of relatively free actors, using the sources of power at their disposal[343] . In the organisational structure that is the social health field of Centre-Cameroon, the 'pluri-actorial' APA is lacking. This frequently leads to situations of political entanglement between health structures at different levels and between those at the same level. In practical terms, two levels of political entanglement can be observed in the Centre's geographical health area: vertical political entanglement (A) and horizontal political entanglement (B).

A- Vertical political entanglement

Fritz SCHARPF spoke of "vertical political entanglement" *(vertikale politikverflechtung)* to describe a context in which the realisation of a   public policy decided at   the "top" depends on the implementation of the policy.
      "depends on
1 cooperative attitude on the part of   inferior levels   which,
      because of their
statutory autonomy, are   critical   players[344] . Gilles MASSARDIER[345] callsa   telphenomenon   "the entanglement of levels of public action'. In the context of this work, vertical political entanglement is understood to mean the existence, within the social health field of Centre-Cameroon, of several decision-making poles to the point where the regional public health delegation (DRSPC) has lost its monopoly on this exercise. We are dealing here with a multi-level public health system requiring a 'degree of vertical coordination'[346] . In this context, when analysing the APAs, the aim is to identify the degree of effective coordination between the authorities and services at the different levels of government[347] . In the absence of this degree of coordination, the result is an "entangled/imbricated or compartmentalised ABS"[348] . This is a constraint on the process of implementing public action.

In the Centre region, health institutions at different levels are constantly overlapping, due to the lack of a coordinated APA, even though the DRSPC has a monopoly on institutional malaria control in this geographical area[349] . In some cases, local players or lower levels adapt the malaria control policy to their own needs, ignoring the decisions taken beforehand or at the same time by players at higher levels. Gilles Massardier is quite clear on this point when he writes: "a modern organisation without uncertainty between the order given at the top of the

---

[343] See CROZIER (M) and FRIEDBERG (E), Idem. p. 92.

[344] Read KUBLER (D) and MAILLARD (J-De), <u>Analyser les politiques publiques</u>. Op. cit... p. 122.

[345] Read MASSARDIER (G), <u>Politiques et actions publiques</u>. Op. cit... p. 205.

[346] This is an internal dimension of the structure of an APA. See KNOEPFEL (P), LARRUE (C), VARONE (F) and SIMARD (J-F), <u>Analyse et pilotage des politiques publiques</u>. Op. cit... p. 244.

[347] Ibid. p. 247.

[348] Ibid. p. 247.

[349] For more details on the monopoly of the Delegation Regionale de la Sante Publique in the institutional fight against malaria, read article 121 (1) and (2) of decree n°2013/093 of 03 April 2013 on the organisation of the Ministry of Public Health in Cameroon.

hierarchy and its application at the bottom is an illusion"[350] . Consequently, despite the 'vertical' nature of the institutional fight against malaria initiated by a key player, political interference or 'overlap' between 'central' and peripheral health structures persists.

The expression "central health structures" in the context of this work should be understood as referring to all the institutions responsible for dealing with malaria throughout the Centre region and whose headquarters are in Yaounde, the regional capital of the Centre. These include the Regional Public Health Delegation of the Centre (DRSPC) and regional and similar hospitals such as the University Hospital Centre of Yaounde (CHUY), the Central Hospital of Yaounde (HCY), the Gyneco-Obstetric and Paediatric Hospital of Yaounde (HGOPY) and the Military Hospital of Yaounde (HMY). The "central" health structures are those with regional competence, the others being considered as "peripheral" or local health structures. The "central" structures represent the top and the "peripheral" structures the base of the health administration pyramid in Central Cameroon.

Political interference or "*overlapping*" occurs in bureaucratic organisations as soon as "multi-factor" APA is lacking. If this is the case, we are faced with situations where "local actors can adapt policy to their own needs, without any real concern for the prior and concomitant decisions of actors at higher levels of government"[351] . The most obvious example of this is the implementation of rapid malaria diagnostic tests (RDTs) in health facilities, which is governed by rationales that have nothing to do with the policy defined by the President of the Republic, Head of State, and which is supposed to be supervised in the region by the Regional Public Health Delegation of the Centre (DRSPC). Since May 2011, RDTs are supposed to be free for children under the age of five, and cost 200 FCFA for patients over the age of five. But what we are seeing is that the health facilities (FOSA) in the social health field in Centre-Cameroon are flouting this prescription by charging an arbitrarily fixed price for RDTs for patients aged under five.

children under the age of five. Worse still, the cost of a RDT varies from one hospital to another for people over the age of five. The words of a woman from Ntui, the mother of a child under the age of five, are quite illustrative: *"Every time I go to the district hospital with my three-year-old daughter, I'm always asked to have her tested for malaria, and it costs ∂a thousand francs. I always pay, I never refuse to pay"*[352] . This social reality is echoed by another native of the town when he says that *"when I go to the district hospital for a malaria test, they ask me to pay 1,000 francs, at the Clinic Le Jourdain, it's 1,500 francs and at the Reference hospital, it's 1,600 francs!*[353] . It can be seen that the health facilities in the Ntui health district enjoy "considerable autonomy of action"[354] in relation to the

---

[350] MASSARDIER (G), <u>Politiques et actions publiques</u>. Paris, A. Colin, 2003, p. 69.
[351] See KNOEPFEL (P), LARRUE (C), VARONE (F) and SAVARD (J-F), Op. cit... p. 248.
[352] Interview with Ms Onanena Bernadette, a native of the town of Ntui on 3 March 2014.
[353] Interview with ATEBA Claude, native of the town of Ntui on 13 March 2014.
[354] See MENY (Y) and THOENIG (J-C), <u>Politiques publiques.</u> Paris, Montchrestien, 1989, p. 237.

DRSPC, which is supposed to coordinate the health structures and monitor disease control programmes. There is therefore an "administrative overlap" between the top and the bottom of the health pyramid structure. The health facilities disregard government regulations and act as they please under the watchful eye of the DRSPC. What is happening in the health facilities in the Ntui district with regard to the cost of a RDT highlights the fact that the vertical nature of the institutional fight against malaria initiated by the DRSPC is being undermined by the health institutions at the base of the pyramid machine, which sometimes act in disregard of the prescriptions issued by the authorities at higher levels. In its own way, the grassroots is active. It escapes the orders and control of the top[355] . Yves MENY and Jean-Claude THOENIG describe the substance of situations of political interference between the top and the bottom of an administrative organisation in the following terms:

"What happens at the base of the pyramid machine is much more complicated than the top thinks. There is in fact a great deal of autonomy of action. The people who do the work, whose behaviour is in principle conditioned by technical and administrative procedures, appropriate them in a discretionary way, to the point of using them as a means of asserting their own internal power in relation to their bosses or to external subcontractors and customers. A complex life of influence games and haggling over the application of rules and procedures takes place at grassroots level, which has little to do with the rationality that the top level claims to manage"[356] .

Between 2010 and 2014, the President of the Republic, Paul Biya, took decisions designed to enable health institutions to respond more effectively to the problem of malaria. He decided to make TDR free of charge for children under the age of five, and set the cost at 200 FCFA for patients over the age of five. This prescription, which was supposed to be implemented by the DRSPC, is instead being appropriated by the health facilities at the base of the Centre's "health pyramid machine". As a result, at the Ntui district hospital, rather than being free, the RDT for children under the age of five costs CFAF 1,000 and varies between CFAF 1,500 and 1,600 for people over the age of five. What is done at the grassroots level of the health pyramid structure in Central Cameroon often has nothing to do with the rationality impelled from the top, because there is always a "strong autonomy of action". Rather than talking about "verticality", it would be better at the moment to talk about "dispersion". In other words, talking about the "vertical" nature of the institutional fight against malaria is utopian. To take it on board is to ignore the "active" nature of the health institutions at the base of the pyramid structure of the social health field in Central Cameroon. In fact, there are always situations of 'overlap' between the top and the bottom, and it is precisely at this level that situations of vertical political entanglement arise, constituting tribulations that are not negligible in the process of intensifying the fight against malaria in the geographical area of Central Cameroon. As a result, this hardening dynamic

---

[355] See MENY (Y) and THOENIG (J-C), Idem, p. 237.
[356] See MENY (Y) and THOENIG (J-C), Idem. p. 237.

becomes "a source of dysfunction"[357] .

In concrete terms, in the social health field of Central Cameroon, the political interference between the central and peripheral health structures is reflected in two major facts. On the one hand, there is the construction of a 'dissonant model of public policy' for the fight against malaria, and on the other, there is collusion between the 'central strategy' and the 'peripheral strategy'.

When we speak of a 'dissonant model of public policy', we are referring to a situation where 'public policy is driven by a set of negotiations and actors that steer it towards peaks or floors that are always different from the initial projects'[358] . Dissonance' in the implementation of a public policy is the immediate consequence of the lack of vertical coordination within an administration. It results in a significant 'gap' between what is intended or decided and what is actually done on the ground. In fact, in the social health field of Central Cameroon, the health structures at the base of the health pyramid tend to consider themselves as autonomous entities in relation to the DRSPC, which is supposed to give impetus at the highest level to the methods of steering public malaria policies. The vertical nature of the fight against malaria is constantly called into question by grassroots players. As a result, these players manage "sources of uncertainty", giving them considerable "room for manoeuvre" in the face of measures imposed from the top by the "key" player, the DRSPC. This is made possible by the policy of state funding of healthcare. In fact, in the Centre region, the finances managed by public health facilities do not come directly from the DRSPC or the district health services (SSD), but from the state. This tends to weaken these "key" health institutions, which do not have control over the financial resources available to the health facilities that are supposed to be subordinate to them. The words of the Head of the Biyem-Assi District Health Service are revealing enough when he says: *"The district health service does not receive any funding from the State; it has a purely administrative role, coordinating activities within the health district. The state allocates funds to health facilities rather than to district health services"*[359] . From this statement, it appears that each health facility at the base of the health pyramid effectively manages a "source of uncertainty" in relation to the District Health Service (DHS), which does not control the exact financial amount allocated to health facilities by the State. As a result, each health facility in a health district has a great deal of "room for manoeuvre", enabling it to undertake initiatives that have nothing to do with what is planned at the top of the pyramid. A power relationship then develops between the top and the bottom of the public health administration. This is precisely where the "overlap" occurs. Empirically, this is reflected in the fact that in some health facilities, such as the *"Jesus Care"* health centre in the Biyem-Assi district, patients who can prove that they are genuinely indigent are

---

[357] Ibid. p. 237.
[358] See EBOKO (F), "L'organisation de la lutte contre le sida au Cameroun: de la verticalite a la dispersion?", Bulletin de l'APAD, 21, 2001. p. 17.
[359] Interview with the Head of the Biyem-Assi District Health Service on 3 June 2021.

treated free of charge. This is purely discretionary power on the part of these facilities, which has nothing to do with what is provided for at the highest level, whether at the DRSPC or the SSD. Rather than being coordinated, the implementation of malaria control policies by health institutions suffers from "administrative overlap".

Furthermore, in the register of political-administrative interference, we should also mention the constant collision between the "central strategy" and the "peripheral strategy" of the fight against malaria. In the social health field of Central Cameroon, the institutional fight against malaria obeys two rationalities: the central rationality and the peripheral rationality. The first is driven by the DRSPC, and the second by the health districts in the social field. We can see that what is wanted at the top of the health pyramid is not what is done at the bottom. This is why we speak of a "collision" between two diametrically opposed strategies. Examples of this are legion. In addition to the fact that RDTs, which are supposed to be free for children under the age of five at "central" level, still have to be paid for at "peripheral" level, we also note that in some health facilities, the treatment of severe malaria for children under the age of five still has to be paid for, contrary to the MINSANTE decision dated 23 June 2014, article nine of which was quoted in the previous chapter. Empirical examples exist in the Ntui and Yoko health districts. At the Ntui district hospital, treatment of simple and severe malaria for children under the age of five is still subject to a charge. Mrs Bernadette Onanena, a local woman from Ntui, explained: *"Every time I go to hospital with my little daughter, I'm always asked to pay. I've never treated her for free"*[360] . In Yoko, according to local people, the price of malaria treatment for children under the age of five varies from one health facility to another. These two examples alone demonstrate that the base of the health pyramid develops a rationality that has nothing to do with that of the top.

At DRSPC level, RDTs and treatment for severe malaria are supposed to be free. But in some health districts they are still charged for! The health facilities therefore benefit from "considerable autonomy of action", which allows them to go against the "central strategy", which is that RDTs and severe malaria treatment for children should be free. It is precisely at this level that the collision between the "central strategy" and the "peripheral strategy" takes place, a concrete translation of the political-administrative interferences. The top does not control the bottom, and the bottom does as it pleases, with verticality giving way to dispersion in the absence of a multi-faceted political-administrative arrangement. This lack of a multi-factorial PAA gives rise to a level of vertical political entanglement that skews the dynamics of institutionalisation of the model fight against malaria by health institutions in the geographical area of Central Cameroon. While the health institutions of Central Cameroon are vertically intertwined, the same scenario can be observed horizontally.

---

[360] Interview with Ms Onanena Bernadette, a native of the town of Ntui on 13 March 2014.

B- Horizontal political entanglement

BENZ and his colleagues spoke of 'horizontal political entanglement' *(horizontale politikverflechtung)[361]* to describe a constellation in which the authorities involved in implementing a public policy are on an equal footing, so that mutual understanding is a *sine qua non for* cooperation. This type of entanglement becomes apparent in administrative organisations when the 'degree of horizontal coordination' is lacking. In this case, the result is not integrated, but horizontally fragmented ABS[362] . Horizontal political entanglement is "manifested much more by a lack of substantial (horizontal) coordination which, in turn, may result from the fact that the actors belong to administrative organisations or to "divergent, or even downright opposing, regional environments because of their respective primary tasks and the interests they represent and defend"[363] .

The definition of the concept of "horizontal political entanglement" that we have adopted for the purposes of this study refers to a horizontal situation in which the health facilities in the health pyramid of Centre-Cameroon are on an equal footing at different levels, to the point where coordination and above all cooperation between them seems impossible if not difficult. It can be seen that, whether at the level of the four regional hospitals, the thirty district hospitals, the seventy district medical centres or the one thousand one hundred and ninety-nine integrated health centres, the health facilities are on an equal footing at their respective levels. As a result, we end up with uncoordinated political and administrative arrangements, characterised by a lack of willingness on the part of the key PLAYERS "to establish the coordination needed to apply the PPA, or even by an explicit strategy of non-cooperation"[364] . The more horizontal the relationships between institutions, the more difficult it is to coordinate them.

Many public policy analysts have shown that horizontality in the process of constructing public action is a factor of complexity or simply of constraint. For Yannis PAPADOPOULOS, AHORIZONTALITY means greater uncertainty as to the behaviour of some and others and the variations in their reactions over time: it brings contingency, which is a central element of complexity"[365] . The political scientist goes on to say that: "horizontal relations pose serious coordination problems, and this particularly affects the political system when it tries to ensure social cohesion in the face of society's centrifugal tendencies"[366] . In the Centre-Cameroon region, horizontal relationships can be seen at health facility level. Horizontal relations between regional hospitals, horizontal relations between district hospitals, horizontal relations between district health centres and horizontal relations between integrated health centres! Worse still, there are no "key" players

---

[361] Read KUBLER (D) and DE MAILLARD (J), <u>Analyser les politiques publiques</u>. Op. cit... p. 122.

[362] Read KNOEPFEL (P), LARRUE (C), VARONE (F) and SAVARD (J-F), <u>Analyse et pilotage des politiques publiques</u>. Op. cit... p. 245.

[363] Ibid. p. 246.

[364] Ibid. p. 246.

[365] See PAPADOPOULOS (Y), <u>Complexite sociale et politiques publiques</u>. Op. cit... p. 36.

[366] Ibid. pp. 36-37.

to ensure the necessary coordination between these different health institutions. As a result, we end up with interactions between two coalitions of health facilities around malaria and a fragmented political and administrative arrangement.

The social health field in Centre-Cameroon is characterised by two coalitions of health facilities, each with opposing strategies. These coalitions have opposing belief systems regarding the cost of treating malaria. On the one hand, there is a coalition of health facilities which believe that the treatment of malaria in children under the age of five should be free of charge, in accordance with government regulations. This coalition includes the regional and similar hospitals and the six district hospitals of Mfoundi. The belief system developed here is that the population is already living in a state of extreme poverty, and the hospitals must help to alleviate the burden on them by opting for free care for children. On the other hand, there is a coalition of health facilities that believe that malaria treatment should be paid for, whether or not the patient is a child. This coalition includes peripheral health facilities such as the district hospitals of Ntui and Yoko in the Mbam-et-Kim department. Here, the hospitals are developing an informal system of patient care, disregarding government regulations. They adopt a strategic approach, examining all possible choices to select those that provide the greatest benefit[367] . The belief system developed here is that the hospital coffers are suffering financially and malaria would be a way of filling this financial void. We can see, therefore, that for the same policy of caring for patients under the age of five, the two coalitions of health facilities develop different rationales and constantly interact. In some hospitals, children under the age of five are actually treated free of charge, while in others, treatment of this social category is still subject to a charge. It is precisely at this level that situations of horizontal political entanglement arise, which undermine the logic of the collective action of those involved in the institutional fight against malaria. These situations of horizontal entanglement arise from the absence of a 'key' actor capable of maintaining coordination between the health centres in the geographical area of Centre-Cameroon.

Each of the two health facility coalitions is defending a very specific cause. On the one hand, the cause being defended is that of the people who are already living in poverty; on the other, the cause being defended is that of the hospitals that are financially affected. This is why, when you visit the health centres in the Biyem-Assi and Cite-verte health districts, for example, you find that the treatment of severe malaria for children under the age of five y is effectively free. On the other hand, when we visit the health facilities in the Ntui district, we find that they still charge for treatment. This is the consequence of a political and administrative arrangement that is not horizontally coordinated. The result is a configuration of highly autonomous players with different interests and perceptions, conducive to

---

[367] Read HALL (P.A) and TAYLOR ROSEMARY (C.R), "La science politique et les trois neo-institutionnalismes", Op. cit... p. 472.

the building of 'coalitions of the willing'. This limits the effectiveness of the model fight against malaria, because it creates a 'split' between the health facilities in the social health field in Centre-Cameroon.

In addition, there is a fragmented political-administrative arrangement between the different public health services in the region, a concrete expression of horizontal political entanglement. In fact, the politico-administrative arrangement is said to be fragmented when an organisation results in "a substantial absence of (horizontal) coordination which, in turn, may result from the fact that the actors belong to divergent administrative organisations, or to "regional milieus", which may even be completely opposed because of their respective primary tasks and the interests they represent and defend"[368] . The more fragmented ABS is within an organisation, the more the interests of the players surface and impose themselves as the normal framework for defining collective action.

The "basic" health institutions in the social health field of Centre-Cameroon are characterised by a high degree of horizontal fragmentation due to the absence of a horizontally coordinated APA. In fact, the integrated health centres in this geographical area are not all on the same footing, which leads each of these health facilities to go it alone rather than coordinating their actions. The words of the general supervisor of the Biyem-Assi district hospital are revealing enough when he says: *"Here in Yaounde, each hospital works alone; we have never been consulted by another hospital about a joint action... We don't know what other hospitals are doing in their health districts"*[369] . If, in the same geographical area, one hospital does not know what is being done in another, this shows that the malaria control APA lacks coordination. As a result, each hospital is pursuing its own interests and acting as it pleases. When four integrated health centres in the region[370] [371] were asked whether children aged zero to five were being treated free of charge, the same answer was given every time: We don*'t know what is done in the other hospitals, so why don't you go and ask them?"* Even though we know that what is said in the hospitals is totally out of step with what is said by the people, who do not recognise that their children under the age of five receive free care in the districts of Ntui and Yoko.

The above statements reflect the lack of verticality and highlight the horizontal nature of relations between health facilities which, instead of co-operating for concerted and joint action, are intertwined horizontally. The result is a fragmented political and administrative arrangement. For something as important as the fight against malaria, it would have been a good idea to establish vertical relations to reduce uncertainties and promote mutual understanding between the health facilities in the Centre's geographical area. Unfortunately, however, there is a

---

[368] See KNOEPFEL (P), LARRUE (C), VARONE (F) and SAVARD (J-F), Loc. cit... p. 246.
[369] Interview with the general supervisor of the Biyem-Assi district hospital on 4 June 2021.
[370] These include the Biyem-Assi, Cite-Verte, Ntui and Yoko district hospitals.
[371] Interview with the general supervisors of the Biyem-Assi, Cite-Verte, Ntui and Yoko district hospitals on 4 and 5 June 2021.

constant lack of cooperation between these facilities.

Paragraph 2: lack of cooperation between health facilities

The term cooperation here refers to collaboration, to mutual agreement between two or more actors engaged in collective action. Cooperating in a bureaucratic administration still means working together on an ongoing basis to achieve consensual, coordinated action. From this point of view, cooperation is a *sine qua non for* successful collective and public action, because reaching an acceptable compromise is the essential precondition for any action[372] . If the malaria problem is to be tackled effectively in the Centre region, the private and public health facilities in the area must cooperate constantly if they are to take rational and organised action against the disease. It should be pointed out that the social health field in Centre-Cameroon is made up of public and private facilities. Apart from the district hospitals, the public sector facilities are in a state of total disrepair and are mostly located in enclave areas. As for private-sector facilities, they are characterised by their cleanliness and modern layout.

In developing what they call *strategic reasoning,* Michel CROZIER and Erhard FRIEDBERG argue that "the participants in an organisation can be regarded as actors, each with its own strategy"[373] . On the basis of this postulate, it is possible to assert that permanent cooperation becomes a *sine qua non* if we want to promote positive coordination within a bureaucratic organisation. However, on observation, the health field in Central Cameroon is characterised by a clear divide between private and public health facilities. Here, we are witnessing a reign of self-centredness, of the 'free ticket' and of sovereign indifference to the problems caused by the options taken[374] . This lack of permanent cooperation between health facilities in the Centre's geographical area is reflected empirically, on the one hand, by the strong dynamic of "self-centredness" of the hospitals (A) and, on the other hand, by the constant formation of "negative coordination" between them **(B).**

A- The strong tendency of health facilities to focus on themselves

The dynamic of "self-centredness" refers to the idea that within an administrative organisation, each segment or sector, before thinking of communicating in any way with the others, goes it alone[375] . It is a real obstacle to negotiation and promotes contingency. It is reflected in the lack of institutionalised transactions in recognised arenas that are the subject of consensus among the participants[376] . Self-centredness" biases the process of strengthening the fight against malaria at institutional level.

In the Centre-Cameroon region, the dynamic of "self-centredness" can be observed at the level of the health facilities, from the top of the pyramid structure downwards, and constitutes a constraint in the process of strengthening the

---

[372] Read CROZIER (M) and FRIEDBERG (E), <u>L'acteur et le systeme</u>. Op. cit... p. 255.
[373] CROZIER (M) and FRIEDBERG (E), Idem. p. 230.
[374] Y. PAPADOPOULOS,Idem. p. 73.
Ibid. p. 73.
[376] Ibid. p. 73.

institutional fight against the malaria problem. In fact, each health facility is "self-centred", i.e. inward-looking, with little concern for other facilities. This "free ticket" approach seriously undermines the process of ongoing cooperation, which is so essential in an organisation that is resolutely focused on collective action.

In the regional and similar hospitals of Centre-Cameroon, there is a strong tendency towards "self-centredness" in the process of implementing public policies to combat malaria. The CHUY, l'HGOPY, l'HCY and l'HMY are characterised by a lack of cooperation and communication. Empirically, this is reflected in the fact that the price of a malaria RDT for adults varies from one hospital sector to another. At the Yaounde Central Hospital (HCY), it costs 200 FCFA, but the price is totally different in other hospitals. At the Hopital Gyneco Obstetrique et Pediatrique de Yaounde (HGOPY) it costs 500 FCFA; at the Centre Hospitalier Universitaire de Yaounde (CHUY) it costs 600 FCFA and at the Hopital Militaire de Yaounde Region 1 it costs 1000 FCFA[377]. This reflects a lack of coordination in the malaria treatment process and of cooperation between these four regional hospitals. In particular, each hospital has no control over the cost of an adult RDT in the other. Each player y goes its own way and sets its price 'discretionarily', a clear illustration of the strong tendency towards 'self-centredness'.

At district hospital level, the tendency to be "self-centred" is even more palpable. Empirically, this is reflected in the fact that the price of treating severe malaria in children under five varies from one hospital to another, and is far from uniform, as each hospital seems to be "sovereign" in carrying out this action. For example, while malaria treatment for children under the age of five is free in the Biyem-Assi and Cite-Verte district hospitals, the cost varies between 2,000 and 4,000 CFA francs in the Ntui and Yoko district hospitals. Similarly, while the RDT for children under five is free at Biyem-Assi district hospital, it costs exactly 1,000 FCFA at Ntui district hospital. This clearly shows that each hospital adopts the "free ticket" strategy, of "self-centredness", enabling it to set the cost of malaria treatment at its own discretion.

At the level of the CMAs and CSIs, there is also a strong tendency towards "self-centredness". Indeed, even at this level of the health pyramid, malaria treatment policies are implemented in isolation. Each CMA and CSI is inward-looking, which is why the cost of malaria treatment varies from one place to another. In the CMAs of Ngambe-Tikar and Mbangassina, malaria treatment (even severe treatment) for children under the age of five is not free, according to the testimonies of many local people. The same social reality can be seen in the Biagnimi and Nditam IHCs. What is curious is that the price of treatment varies from one CMA to another and from one CSI to another. On average, it is 2,500 FCFA in the CMAs and 4,000 FCFA in the CSIs.

This strong tendency towards "self-centredness" on the part of the health facilities

---

[377] These costs were obtained from the various hospitals and from local people after an interview carried out on 15, 16 and 17 June 2021.

biases the process of tightening up malaria treatment because it leads to a lack of compromise on the actual ways in which malaria is managed in the health social space of Central Cameroon. The fact, for example, that the Biyem-Assi district hospital "does not know what the other hospitals are doing in their health districts" shows that the fight against malaria, which causes so many deaths in the Centre region, is still in its "embryonic phase" from the point of view of the health institutions supposed to be responsible for this fight. It is no doubt this lack of institutional cooperation that led Yannis PAPADOPOULOS to write that "common sense frequently attributes bureaucratic inconsistencies to a lack of coordination between the various administrative bodies"[378] . However, a good malaria treatment policy requires the creation of bodies designed to encourage negotiation and cooperation, while helping to make them visible (round tables, ethics committees, consultation forums, etc.)[379] . Without this, the result will be "negative coordination".

B- Negative coordination between public and private health facilities

Negative coordination occurs when the actors involved in implementing a public policy achieve minimal results in terms of gains in social well-being. This is due to the presence of a large number of players who have to take part in negotiated decisions. This complicates the decision-making system and is likely to increase the risk of deadlock, or at the very least greatly reduce the capacity to take decisions that depart from the *status quo*[380] . Negative coordination is the concrete expression of the failure of joint decision-making and positive coordination[381] . The consequence of negative coordination is to generate measures that are "pareto-optimal"[382] .

In the context of this work, negative coordination is taken to mean the administrative situation in which players have to come to terms with each other in order to reach coordinated decisions, but find it difficult to do so because of the conflicting interests they share. It is the concrete expression of the failure of positive coordination and constitutes an institutional constraint on the process of implementing public policies. In the Centre-Cameroon region, there is negative coordination between the actions of public and private health facilities. In fact, for the same policy of caring for malaria patients, different rationales can be observed,

---

[378] PAPADOPOULOS (Y), <u>Complexite sociale et politiques publiques</u>. Op. cit... p. 98.

[379] Ibid. p. 73.

[380] Read KUBLER (D) and DE MAILLARD (J), <u>Analyser les politiques publiques</u>. Op. cit... p. 125.

[381] We speak of positive *coordination* when the players involved in implementing a public action are able to deliberate on the best response to a problem while simultaneously dealing with issues of distribution among them of the costs and benefits associated with an action. See KUBLER (D) and DE MAILLARD (J), Idem, p. 124.

[382] The concept is named after the Italian economist Vilfredo Pareto, who used it to describe a state of society in which the well-being of one individual cannot be improved without damaging that of another. It is thus opposed to the concept of the "Kaldor-Hicks optimum". In the context of public policy analysis, this concept is much more widely used by theorists of actor-centred institutionalism to describe a context in which a multitude of actors with divergent interests prevent the implementation of a joint decision precisely because none of these actors will give in for fear of losing out. Each player wants to retain its room for manoeuvre and does not intend to lose it at the instigation of another player. As a result, all we end up with is negative coordination within the administration and minimal decisions in terms of potential gains in social welfare.

leading to a high degree of negative coordination. While in some public-sector health facilities, particularly those in the city of Yaounde, malaria treatment for children under five is free, it is charged for in private-sector facilities. Similarly, while intermittent preventive treatment (IPT) based on sulphadoxine pyrimetamine for pregnant women is free in public sector health facilities, it is charged for in private sector facilities. It is as if each sector of the bureaucratic health system is inward-looking and has no intention of communicating in any way with the other. The main reason for this state of affairs is that public-sector health facilities receive substantial subsidies from the state, unlike private-sector facilities, which rely solely on their own financial resources to carry out their activities. For private-sector health facilities, the logic is simple: "each patient contributes in his or her own way to promoting the development of each private facility"[383] . It follows from this that the money collected on the treatment of children under five and on IPT for pregnant women contributes to the development of private health facilities and justifies the way they are set up. The objectives or preferences of these facilities are then defined in a way that is external to the institutional analysis[384] . They seek to maximise their success in relation to a set of objectives defined by a given preference function[385] . The consequence of this 'dissonance' between the public and private sectors is the 'saturation' of public health facilities in terms of the number of people to be treated per day. In other words, people, the vast majority of whom are indigent, are turning much more towards the public sector, to the detriment of the private sector, which is too expensive! This has repercussions on the quality of patient care and considerably limits the effectiveness of the fight against malaria in the region. In the district hospitals of Biyem-Assi and Cite-Verte, at the entrance to the doctor's office, there are long bench-tables on which many patients are sitting waiting for consultations and treatment. The risk is that some patients won't be able to get treatment in time, and the worst can happen. However, in private hospitals such as the "jesuscare" health centre and the "Dispensaire Catholique Notre Dame de la Merci", the situation is completely the opposite: the benches are sometimes empty. This socio-institutional reality is the result of a lack of cooperation and therefore of negative coordination between the public and private structures in the social health field in Central Cameroon. Here, the health facilities in each sector are called upon to communicate better, to cooperate if we want to limit the cases of 'overload' suffered by the public health facilities. The private health facilities must come to the "rescue" of the public health facilities, which are suffering from an overflow of sick people. This is only plausible if there is positive co-ordination initiated by a "key" player.

In short, in the Centre's social and health field, the initial setbacks to the process of institutionalising the fight against malaria are directly linked to the health

---

[383] Interview with the general supervisor of the Notre Dame de la Merci Catholic dispensary on 4 June 2021.
[384] Read HALL (P.A) and TAYLOR ROSEMARY (C.R), "La science politique et les trois neo-institutionnalismes", Op. cit... p 472.
[385] Ibid.

institutions. They are an obstacle to the construction and implementation of the decisions of the central public authorities (President of the Republic, and MINSANTE) which are supposed to be implemented at all local levels y including the Centre region. These so-called institutional tribulations stem from political entanglement and lack of cooperation between health institutions. In fact, the institutional structures responsible for combating malaria are first and foremost politically entangled, both vertically and horizontally. As far as vertical political interlocking is concerned, despite the willingness of the DRSPC, the 'key' player in the fight against malaria, to give impetus to a top-down approach, there is always political and administrative interference, or a lack of cooperation between the various institutions.

"overlaps' between central and peripheral structures, between the top and the bottom. The process of horizontal political entanglement can be seen at the base of the Centre's bureaucratic health organisation. Being on an equal footing and quasi-autonomous, the health facilities are horizontally intertwined. Empirically, this is reflected in the construction of two coalitions of health facilities around the problem of malaria, and in a fragmented political and administrative arrangement of the health services.

Furthermore, the institutional structures responsible for combating malaria are characterised by a lack of cooperation, which is another institutional constraint. In fact, in the social health field of Centre-Cameroon, we observe on the one hand that there is a strong "self-centred" dynamic in the health facilities. This is made possible by the lack of cooperation between hospitals and the constant formation of negative coordination. This unfortunate reality only serves to skew the process of positive co-ordination, which aims to rationalise the implementation of a policy. On the other hand, there is a glaring lack of joint action, which leads to the complexity of joint hospital action and the non-existence of mutual partisan adjustment. As a result, we end up with an implementation that operates according to the "debrouillardise model". While the health institutions are the tribulations in the process of toughening up the model fight against the malaria problem, it should be noted that, in the course of this process, other, much more serious constraints are also distorting this dynamic of toughening up in the Centre region. This time, we are talking about tribulations directly linked to the social field or the territory in which the fight against malaria is institutionalised. We will no longer speak of 'institutional complexity', but of 'territorial and social complexity'.

SECTION II

The tribulations Пёез to the field of deployment of actors: the "territorial and social complexity"

The social field in which the fight against malaria is institutionalised is also a source of tribulations, since this field is 'complex'. The complexity here is no longer due to the institutions but to the socio-cultural and territorial space of Central

Cameroon itself. As a result, we no longer speak of 'institutional acomplexity' but of 'territorial and social acomplexity'. According to Yannis PAPADOPOULOS[386] , 'social acomplexity' arises when, during the formulation or implementation of a decision, a considerable number of actors with different interests, priorities and logics intervene at one time or another. In the context of this work, the notion of 'social complexity' goes beyond Professor PAPADOPOULOS' definition, which reduces it to the presence of a multitude of players with divergent interests. However, it also translates into the presence of a social space or field which in itself possesses constraints likely to bias the hardening of the model fight against the malaria problem. Taken in this sense, 'social acomplexity' will be lost here as a formidable limitation on state action"[387] and social actors. Territorial complexity" is understood here as any difficulty inherent in the territory that hampers the action of social actors in the social field.

The Centre-Cameroon region is an area with a high degree of 'autonomy' which is experiencing socio-anthropological realities which other areas are not experiencing. It is in fact a complex territory if we take into account its architecture, its surface area, the number of its inhabitants and the prevailing social *habitus*. These socio-spatial realities are all constraints on public action to combat malaria. In concrete terms, in this social health field, the tribulations of the dynamics of institutionalisation of the fight against the malaria problem are directly linked to the territory, the 'aresstissants' and their *habitus*. Thus, the following lines will be devoted to an analysis of the socio-cultural and territorial tribulations that skew the dynamics of the hardening of the modeled fight against the malaria problem in the Centre region. These tribulations are linked, on the one hand, to the territory **(paragraph 1)** and, on the other hand, to the nationals and the "apolitical" drug policy they adopt **(paragraph 2).**

Paragraph 1: The tribulations of territorial complexity

Maryvonne Le BERRE defines territory as "the portion of the earth's surface appropriated by a social group to ensure its reproduction and the satisfaction of its vital needs. It is a spatial entity, the place where the group lives, inseparable from it"[388] . It is clear from this definition that the territory is not always linked to geographical and administrative data, but is rather a construct of human action. It is individuals who shape what we will call 'their territory' simply by settling there and carrying out social activities. Thierry PAQUOT argues that "territory is the result of human action, not the result of relief or physical and climatic conditions"[389] .

For Gregoire FEYT, "by territory, we must certainly understand an administrative and functional reality, but it still often remains an abstract space at the cognitive and affective level, i.e. capable of being constructed by the actors"[390] . FEYT's

---

[386] See PAPADOPOULOS (Y), <u>Complexite sociale et politiques publiques</u>. Op. cit... p. 55.
[387] Ibid. p. 17.
[388] Citee par PAQUOT (T), " Qu'est-ce qu'un territoire"? In <u>Vie Sociale</u>. №2, 2011, pp. 23-32.
[389] Read PAQUOT (T), Idem.
[390] Read FEYT (G), "Redistribution des pouvoirs, redistribution des cartes. La connaissance des territoires, enjeu

definition is the one that suits the conception of 'territory' adopted in this work. Thus, the territory refers both to an administrative reality obeying the administrative division of a State and to an abstract data, constructed by the members who compose it.

To say that a territory is complex is to admit that, geographically and structurally, it is characterised by 'anomalies' likely to hinder the activities of the individuals who live there. This prompts the researcher to analyse territorial policies, and the issue of territorialising public policies lies at the heart of this reflection[391] .

The social health field of Central Cameroon is characterised by "functional differentiation", i.e. it has its own tasks, its own "coding" and its own "territorial logics" which constitute its particularity. In other words, it has socio-spatial, socio-geographical and physical-climatic realities that make it a particular sub-system different from neighbouring sub-systems. This territorial particularism has led to tribulations that have distorted the dynamics of institutionalising the fight against malaria. These include the socio-territorial realities, such as the poor state of the roads, the isolation of the area, the vastness of the territory and the ecologically favourable areas. Since the Centre region obeys a Yaounde/periphery dialectic, as demonstrated in the preceding chapters, the analysis of the tribulations of the fight against malaria linked to territorial complexity must here be carried out under the prism of this socio-geographical reality. In other words, in the geographical area of Central Cameroon, the 'territorial tribulations' of the dynamics of the institutionalization of the modal fight against malaria are not experienced in the same way everywhere. They differ according to whether one finds oneself in the health districts of rural areas **(A)** or in those of Yaounde **(B).**

A- The tribulations of territorial complexity in rural districts

In the Centre region, the rural districts are those outside the city of Yaounde. There are twenty-four (24) of them and their configurations are virtually identical. They differ greatly from the districts of the city of Yaounde and present particular socio-spatial and socio-anthropological realities. As a result, the trials and tribulations of the tougher approach to malaria control in these areas are not the same as those experienced in Yaounde. Here, two socio-territorial realities influence the activities of the implementers. The hostility and isolation of the area on the one hand (1) and the presence of medical deserts on the other **(2).**

1. The tribulations of hostility and isolation

The health sub-systems in the rural areas of the Centre are hostile and highly enclosed in places, so that they become restrictive for the social fight against malaria. In these areas, the process of intensifying the fight against malaria by a range of social actors suffers from certain territorial distortions. Firstly, these areas are hostile in places. In fact, to reach certain rural health districts, you have to walk through an entire forest and face certain insects that are harmful to peace and

---

inedit de 1 action publique" in FAURE (A) and NEGRIER (E), Op. cit... p. 133.
[391] Read BERTHET (T), Op. cit... p. 44.

health. This is the case in the Ntui and Yoko health districts in the region's largest and, curiously, most hostile department. In these districts, the "restricted" distribution of Mil by the state to pregnant women between 2003 and 2005 was a real fiasco. The main reason for this was that the community health workers (CHWs) in charge of the distribution process faced hostile sub-systems, so they were unable to cover three-quarters of the households with pregnant women living y.

Territorial hostility is reflected in the fact that to enter the Mbam-et-Kim department from the town of Obala in Lekie and reach Ntui, you have to pass through a forest inhabited by insects that are harmful to the skin, known as "mout-mout"[392] . Some community health workers from Yaounde admit that they were unable to enter the Ntui health district for fear of being bitten by the midges[393] during the "restricted" distribution of Mil to pregnant women between 2003 and 2005 and during the household enumeration between 18 and 30 April 2016. The presence of midges at the entrance to the town forced some CHWs to turn back, thus reducing the number of actors in charge of the millet distribution and household enumeration process. Even those CHWs living in the Ntui district admitted that they had not been able to reach the health areas located in densely vegetated areas. Pregnant women living in the "Ngoro" and "Ndjame" health areas hardly felt the presence of the CHWs in 2005, and consequently did not receive a Mil. Similarly, the households in these sub-systems were not counted in 2016 and consequently did not receive any coupons or LLINs.

The health district of Yoko presents almost similar realities, as we also find the presence of harmful insects such as midges. Moreover, in this sub-system, there is a strong presence of so-called "wild" wasps, whose bites leave bodily lesions on their victims. There are also large numbers of "tick fleas"[394] . These "wild" insects do not encourage the presence of humans in this environment unless you are used to being around them. This acts as a brake on the drive to step up the fight against malaria.

It is therefore easy to understand why the distribution of mosquito nets has always ended in failure in these outlying areas. The social players responsible for stepping up the fight against malaria are finding it extremely difficult to operate in these hostile geographical areas. This state of affairs goes some way to explaining why, out of 129,782 pregnant women registered in the Centre region in 2003, only 3,406 Mil were distributed to those living in rural districts, giving a coverage rate of 3.2%. This also explains why, during the major LLIN distribution campaign in 2016, the 24 outlying districts received only 878,115 LLINs, while the six Yaounde districts received 1,432,511!

---

[392] This is a purely indexical expression whose meaning can only be local. In the Ntui subsystem, this expression is used to designate "midges", which are small black insects that attack the skin of humans and suck their blood. They also have the characteristics of mosquitoes and can be the source of many diseases.
[393] This was entrusted to me by Mr Nga Sylvestre, who was a CHW in the Ntui health district during the distribution of LLINs in 2011.

Districts in rural areas are also a brake on the fight against malaria because they are highly isolated. In these sub-systems, there is a defective road network that is unsuitable for traffic. To get to the Ntui health district from Yaounde, for example, is a real ordeal, as a community health worker told us:

"We had the opportunity to experience this ordeal in 2003, 2004 and 2005. First of all, to get to the district from Obala, you had to take the Batchenga-Nachtigal stretch, which is the gateway to the town of Ntui. This stretch was a real ordeal in both the dry and rainy seasons. It's an unpaved track, full of potholes. In the dry season, the dust is in full swing and it's impossible not to suck it up. In the rainy season, it's only death! y there are quagmires everywhere, puddles and lots of mud. Whether you're on foot or in a car, you'll be

---

[394] These are small, black insects that live in the soil and can penetrate human feet without causing the slightest pain. Their presence in the foot causes the toes to itch, swell and cause pain.

was having a hell of a time. Many y had to turn back before even reaching Nachtigal. Those lucky enough to make it to Nachtigal faced a new ordeal as they had to cross the Sanaga to reach Ntui. The crossing was always made by ferry, which has the particularity of holding only a limited number of people and travelling very slowly. It often broke down, and when it did, they had to reach the Ntui district by pirogue, with all the risks that entailed".

These comments highlight the constraint posed by the lack of communication routes in the outlying districts. The MINSANTE agents from Yaounde, responsible for distributing the Mil, were confronted with this territorial constraint to such an extent that some of them had to abandon the process along the way. Those who managed to penetrate these rural districts fought a titanic battle. Once inside these sub-systems, the ordeal persists, if we take into account the words of this CSA:

"The agents responsible for distributing the nets who even lived in Ntui also went through an ordeal. There are certain health areas that are impassable when it rains. It's impossible to get to Ndimi, Ngoro or Talba without taking risks. The road is slippery, and there are puddles everywhere, with the possibility of a landslide! There's also forest everywhere, as this is a forested area. The risk of being bitten by a snake could not be ruled out"[396].

It is therefore easy to understand why certain health areas in rural areas such as "Ngoro" and "Ndjole" did not receive a single mosquito net from the State between 2003 and 2005, simply because the CHWs were unable to reach these areas because they were so isolated. In his work on the "Akonolinga health district", Raphael OKALLA lays bare the difficulties of reaching the Akak health area via the road network when he writes: *"The village of Akak is located some forty kilometres from the Endom hospital. The road is so bad in the rainy season that it takes three hours by all-terrain vehicle to reach it from Endom"*[397]. Hostility and the high degree of isolation in rural districts also skewed the process of distributing LLINs during the mass campaign in April-May 2016. For this reason, the results recorded in these sub-systems were mediocre in terms of population coverage with LLINs[394].

In addition, the hostility and isolation of the outlying areas also affected the work

---

[394] Interview with Mr Nga Sylvestre, ASC of the Ntui health district in 2011, on 14 June 2021.
[395] Interview. Ibid.
[396] Read OKALLA (R), "Le district de sante d'Akonolinga", in Bulletin de l'APAD, 2001, p. 5.
[394] See table 19 above.

that APP community health workers wanted to carry out as part of the policy known as "home-based management of uncomplicated malaria" (PECADOM) during the "*LLIN Hang-Up*" campaign between 02 and 14 January 2011. These were mobile campaigns to detect cases of uncomplicated malaria and treat them at home, inspired by Eugene JAMOT's strategy, while at the same time distributing LLINs. The CHWs in charge of this operation encountered difficulties in implementing it. The hostility in the area meant that they were unable to visit households in rural districts. For example, in the Ntui district, the CHWs limited the operation to the town centre. This is reflected in the observation that only people living in the "Ntui" health area have гсси APP CHWs in their homes and know what PECADOM is all about. Households living in the eleven other health areas (Biakoa, Mbangassina, Ndimi, Ndjame, Ngambe Tikar, Ngoro, Nguila, Nyamanga II, Nyamoko, Talba and Voundou) were not made aware of the programme and their patients were not treated at home because the CHWs who were supposed to carry out these social actions were all based in the "Ntui" health area located in the centre of the town. This has limited PECADOM's action, which appears to the people living in the hostile, enclave health areas to be "just a slogan with no real content"[395] . Apart from the hostility of the environment and its high degree of isolation, another territorial constraint severely limits the effectiveness of the intensified social fight against malaria in the rural health districts of Central Cameroon: the presence of "medical deserts".

2. The tribulations of "medical deserts

The term "medical deserts"[396] refers to the difficulty people have in gaining geographical access to healthcare centres. It highlights the problem of geographical access to healthcare for people from a particular social group, which is reflected in the distance of health centres from their homes. According to WHO indicators, the standard for geographical access to healthcare is five (5) kilometres and one health facility for every ten thousand (10,000) inhabitants[397] . However, according to MINSANTE, "there is an infrastructure imbalance between the different regions and even between health districts, where some people still live more than 20 km from a health facility"[398] . In Cameroon, it has been noted that in very large rural areas, many people still have to travel long distances to find a health centre. As a result, the primary healthcare strategy is difficult to implement for many people[399] . On observation, this observation is true for the rural health districts of Centre-Cameroon because many people living in these areas are confronted with the presence of medical deserts and have difficulty accessing health centres. This state of affairs is a serious obstacle to stepping up the social fight against malaria.

---

[395] Read SIMBE AVORE (T.D), <u>Les politiques publiques de traitement du paludisme dans l'arrondissement de Mbouda (1952-2014)</u>. Memoire de master en science politique, Universite de Dschang, 2016, p. 84.
[396] Read FOE NDI (C), <u>La mise en auvre du droit a la sante au Cameroun</u>. These de doctoral en Droit public, Universite d'Avignon, 2019, pp. 196-198.
[397] See FOE NDI (C), Idem. p. 198.
[398] Quoted by FOE NDI (C), Idem. p. 197.
[399] Ibid. p. 198.

The rural health sub-systems of Central Cameroon are faced with "medical deserts", which are expressed empirically either by a "vacuum" in terms of health centres, or by the absence of quality health facilities able to provide health care for the population. As a result, thousands of people have difficulty accessing a quality health centre in the event of illness. This is the case in the village of "Ndimi" in the Ntui health district, which has a population of around 1,500 but no quality health facilities. Those that do exist have no technical facilities and are in a state of total disrepair (this is the case of the Ndimi and Ehondo IHCs). What's more, they are facing a glaring shortage of hospital staff, because the doctors assigned to these areas are refusing to take up their duties and are going to live in Ntui or even Yaounde. The same applies to the "Ngoro" health area, which has a population of 13,892. The consequence of this is that people are sometimes obliged to migrate to the district hospital in Ntui in the event of illness for those who seek conventional medicine. However, the "Ndimi" and "Ngoro" health areas are located some twenty kilometres from the Ntui district hospital, well above WHO standards. How is it possible to envisage any ownership of their health by the members of a community who are obliged to travel twenty (20) or even eighty (80) kilometres to find a health centre? asked Mr FOE NDI intelligently at[400] . What's more, in the rainy season, it takes at least two hours and two thousand francs to get to the district hospital in Ntui. The same reality can be observed in the Akonolinga health district, where Raphael OKALLA shows that the village of Akak has no quality hospitals, and as a result, the people who live there are obliged to travel some forty kilometres to reach the Endom hospital on a road that is "so bad in the rainy season that it takes three hours with an all-terrain vehicle to reach it from Endom"[401] . In the Yoko health district, there are areas where there are virtually no hospitals. This is the case in the village of Mfoumbe, which has almost 1,000 inhabitants but no health centre. Here, people are forced to travel almost thirty (30) kilometres to reach the nearest integrated health centre.

The above is ample proof that people living in rural areas of the Centre region have difficulty accessing healthcare. This is where the concept of "medical deserts" comes into its own. In other words, in this part of the region, there are social areas where there are virtually no hospitals, forcing local people to travel miles to see a doctor worthy of the name. This state of affairs encourages the emergence of parallel malaria treatment facilities and undermines the effectiveness of the social fight against this disease.

In short, in rural health districts, the institutionalisation of the fight against malaria is influenced by tribulations directly linked to the territory in which it takes place: this is "territorial complexity". These tribulations result from the hostility of the environment, its isolation and the presence of medical deserts which make the social fight against the malaria problem derisory. The districts located in the city of

---

[400] Ibid. p. 198.
[401] See OKALLA (R), Idem. p. 5.

Yaounde are not immune to the vagaries of the territory. They too have limitations directly linked to territorial complexity, which skew the dynamics of the fight against malaria.

B- The tribulations of territorial complexity in the Yaounde districts

In the social and health field of Central Cameroon, the constraints linked to the territory that hinder the dynamics of institutionalising the fight against malaria are not only visible in rural areas. The city of Yaounde also presents territorial realities that make the fight against the malaria problem more complex, even though they differ from those experienced on the outskirts. It should be remembered that the city of Yaounde has six (6) health districts with a virtually identical morphology as a result of the fact that they are all located in an urban area. Here, the process of implementing public action on malaria has not been easy for the various social actors. They had to deal with a territory with a 'particularism' and a 'functional differentiation' that territorialized the fight against the disease in this geographical area. Two tribulations linked to territorial complexity can be identified in the Yaounde sub-systems: the predominance of 'ecologically favourable areas' (1) and the immensity of the health districts (2).

1. The trials and tribulations of "ecologically favourable zones

In his work on malaria control in urban areas, Mr MEVA'A ABOMO Dominique draws a distinction between what he calls "favourable ecological settings" and "unfavourable ecological settings". The former are areas at high risk of malaria transmission, where malaria exposure levels are consequently higher[402] . The latter are areas where the risk of malaria transmission is low, insofar as the vectors still do not find the useful ecological conditions necessary for their reproduction and development[407] .

The term "ecologically favourable areas" refers to districts where socio-geographical conditions are conducive to the development of malaria vectors. They are characterised by the existence of certain sites such as water corns, river beds, low-lying wetlands, urban wastelands, areas of stagnant water, unkempt areas, etc.[408] . The vast majority of Yaounde's health districts are characterised by the presence of this type of area, which makes it difficult to combat malaria in society. Although they are inherent to the configuration of ecosystems, they are also the result of human activity.

---

[402] Read MEVA'A ABOMO (D), "Le fardeau de la lutte contre le paludisme urbain au Cameroun : etat des lieux, contraintes et perspectives", Op. cit... p. 31.

407 Ibid. p. 31.

408 Ibid. p. 31.

409 This is an indexical expression whose meaning is purely local. Elie describes the Biyem-Assi district as the most insalubrious because of the presence of untended low-lying wetlands. The local people refer to it as a "swim-slide" because of its accident-prone geographical configuration, which is not only slippery but can also lead to swimming for those who find themselves in it.

Several areas of Yaounde are ecologically favourable for the spread of malaria. They are found in all the health districts, and are characterised by low-lying, marshy areas, riverbeds and untended areas of stagnant water. In the "Biyem-Assi" district, the area known as "nage-glisse"[409] is an ecologically favourable area par excellence. It is a vast area of wasteland, unhealthy and full of urban wasteland. It's a prime area for the development and spread of mosquitoes. Here, during the day and at night, mosquitoes can be heard hissing, exposing many people to the risk of malaria. The so-called "swim-slide" area needs to be maintained by the public authorities, but households can do nothing about it, given its size, depth and the risk of landslides. In the Elig-Edzoa district, there is a similar area known as "marche Elig-Edzoa bas-fond". This area is characterised by the use of marecages that have been left abandoned for ages, exposing the many vendors there to mosquitoes. In the Etoa-Meki district, the area known as "Ntaba" is a real ecological hot spot. The area is mainly made up of urban wasteland, wasteland and areas of stagnant water. The Yaounde authorities had banned the construction of houses in this area, but because of poverty and corruption, many families continue to live there, exposing themselves to the risk of malaria transmission by mosquitoes. The "Tongolo", "Mballa II" and "Mfandena" neighbourhoods are also ecologically favourable because of the high number of marecages found there. These marecages are sometimes located in the vicinity of residential houses.

Favourable ecological zones" are found in large numbers in the city of Yaounde, much more so than in rural areas. These areas are ideal for the spread of malaria because of the high proportion of mosquitoes found there. In this respect, they constitute a real constraint on the dynamics of institutionalising the model fight against malaria, because they practically annihilate the action of social actors. The various distribution sequences for millet and LLINs become pointless if mosquitoes can 'roam' in the middle of the day and outside homes. Another constraint on the construction of a model malaria control system remains in the city of Yaounde. This is the vastness of its health districts.

2. The tribulations of the vastness of health districts

The six health districts of Yaounde are characterised by the fact that they are vast in relation to the neighbourhoods converted by these districts[403] . The Ndjoungolo health district covers the following neighbourhoods:

- Bastos;
- Elig-Edzoa;
- **Elig-Essono ;**
- **Emana;**
- **Essos;**
- **Etoa-Meki;**
- Etoudi;

[403] The neighbourhoods in bold have been designated as health areas in the Centre region. The other neighbourhoods have not been designated as health areas and are not included in this list.

- Mango tree;
- **Mballa II;**
- **Mballa V;**
- Messassi;
- **Mvog-Ada ;**
- **Nfandena ;**
- Ngousso ;
- **Nkomesseng ;**
- **Nkolondom ;**
- **Nlongkak;**
- Nyom;
- Omnisport;
- Olembe;
- Titi-Garage;
- **Tsinga-Village**
- Tongolo ;

The Biyem-Assi health district covers the following neighbourhoods:
- **Akok Doe;**
- **Biscuits;**
- **Biyem-Assi 1;**
- **Biyem-Assi 2 ;**
- Camp Yeyap ;
- **Etoug-Ebe;**
- **Melen;**
- **Mendong;**
- **Mvog-Betsi;**
- **Nkolbikokl;**
- **Simbock.**

The Nkolndongo health district covers the following neighbourhoods:
- Biteng;
- **Ekounou ;**
- Elig-Mbida;
- **Kondengui;**
- **Meyo ;**
- Massamendongo;
- **Mimboman 1 ;**
- **Mimboman 2 ;**
- Mvan;
- **Nkolndongo 1**
- **Nkolndongo 2**
- **Nkomo ;**
- **Odza.**

The Cite-Verte health district covers the following neighbourhoods:
- **Brickworks;**
- **Carriere;**
- **Cite-verte;**
- **Ekoudou ;**
- Madagascar;
- **Messa;**
- **Mokolo ;**
- **Nkolbisson;**
- **Nkomkana;**
- **Tsinga;**
- Tsinga-Elobi;
- **Tsinga-Oliga.**

The Efoulan health district covers the following neighbourhoods:
- **Afane-Oyoa ;**
- **Ahala;**
- **Etonian;**
- Mvolye;
- **Ngoaekelle;**
- **Nsimeyong;**
- **Obili;**
- Obobogo.

The Nkolbisson health district covers the following neighbourhoods:
- **Ekorezock;**
- **Etetak;**
- Nkol-Afeme ;
- **Nkolbisson;**
- **Nkolnkoumou ;**
- **NkolsoO**
- **Nkolso Oyomabang;**
- **Name Nnam ;**
- **Oyom-Abang.**

Yaounde's six health districts each cover an area of between 20 and 50 km2, which is why they can be described as immense. This vastness is a significant constraint on the fight against malaria. As proof of this, the household enumeration campaign in the city of Yaounde between 18 and 20 April 2016 was difficult to carry out in the social field because of the vastness of Yaounde's sub-systems. This was the case in the Biyem-Assi health district, which covers two Yaounde arrondissements, Yaounde III and Yaounde VI. In this district, the MINSANTE agents in charge of the enumeration operation were faced with a very vast territory, if we take into account the following comments made by one of these agents: *"The household enumeration campaign we conducted in 2016 was not a success overall, because we were unable to cover certain neighbourhoods completely. This was because these neighbourhoods were vast and we only had 288 health workers trained and sent out into the field*"[404] . So how do you cover six health sub-systems ranging in size from 20 to 50 km2 when there are only 288 health workers? This question led to the idea that the 2016 household enumeration campaign was destined to fail. For such an operation to bear fruit, you need a considerable number of agents capable of covering the 180 km2 of Yaounde without the slightest difficulty. But this requires considerable financial resources, which the State does not yet have. From this point of view, the vastness of the health districts was an obstacle to the household enumeration policy. This is why, out of a sample of 1,006 households in the city of Yaounde, only 628 had received a coupon between 18-30 April 2016[405] . In total, 32.7% of households in Yaounde did not receive a coupon/ticket in April 2016[406] .

Similarly, the campaigns to distribute Mil to pregnant women in the Centre's social health field from 2003 to 2005 were also affected by the territorialisation of Yaounde's districts. The sheer size of these districts meant that the expected 60% coverage rate could not be achieved. In 2003, the coverage rate was 6.5%, in 2004 16.5% and in 2005 31.9%. These rates are well below average. One of the reasons given by the CHWs in charge of distributing these Mil is that they had to deal with districts that were difficult to navigate because of their vastness. The Biyem-Assi and Ndjoungolo health districts, reputed to be the largest in Yaounde (with around 300,000 inhabitants in 2019, according to the administrative authorities), exhausted the CHWs because they had to go round all the households registered at the market. This seriously hampered the process of distributing Mils to pregnant women.

In short, the social health field in Central Cameroon is experiencing territorial tribulations which are hindering the process of institutionalising the fight against malaria in this area. These tribulations can be observed in both rural and Yaounde health districts. In the periphery, hostility, isolation and the presence of medical deserts are all tribulations linked to territorial complexity, which distort the process of strengthening the social fight against the malaria problem. In the city of Yaounde, the territorial tribulations revolve around the presence of ecologically favourable areas and the vastness of the health districts. This 'territorial complexity' is not conducive to the deployment of the social actors responsible for steering social policies. Aside from these territorial tribulations, the implementation of public action to combat malaria comes up against other difficulties. These are directly linked to the "nationals" and the prevailing drug policy.

Paragraph 2: The tribulations of social complexity: the proliferation of burdens inherent in "nationals" and "medical treatment

In the geographical area of the Centre, the individuals towards whom the

---

[404] Interview with a malaria no more agent who did not wish to reveal his identity.

[405] See the document entitled <u>Cameroon: enquete post campagne sur l'utilisation des moustiquares impregnees d'insecticide a longue duree d'action 2016/2017</u>. Op. cit... P. 73.

[406] Ibid.

implementation of public malaria policies is directed also constitute tribulations to the hardening of the modeme fight against this disease. Their characteristics and lifestyles are not always conducive to receiving the   action of the social actors in charge of steering social policies. The "primarian frameworks"[407] and the "taken-for-grantedness"[408] shared by these people   are obstacles to the deployment of implementers,   especially when the latter are   not "affiliated" to the social partners.

The ethnomethodological   current sees affiliation as a process which consists in discovering and appropriating the routines and self-evident ways - ethnomethods - hidden in the practices of a society, without which the novice will not be able to join his new group, and will quickly be in a situation of abandonment[409] . In other words, to join a group is still to become a member of it. Implementing social policies to combat malaria in a given area first imposes the imperative duty to join the social group on which you wish to act and bring about change. Failing this, local people will always appear to be an obstacle to the deployment of social action, precisely because the implementers are not in tune with the world around them. In the social and health field of Central Cameroon, the nationals (ethnomethodologists prefer the notion of *members)* can be classified into two main groups: those living in urban areas and those living in the periphery. Within these two main groups, there are also sub-members belonging to different health districts. The members of each sub-system share common values, cultures, codes and frameworks that make them individuals in their own right, different from the members of the other sub-systems. The result is a veritable "social complexity" that must be managed by the social players responsible for stepping up the fight against malaria. The way of life of the local people and the cognitive scripts they share constitute constraints to the dynamics of institutionalising modern malaria control in the Centre region. Another constraint prevails in this geographical area. It is directly linked to the drugs in circulation. There is a 'crisis' in anti-malarial drugs in the region, and most of those that do exist are counterfeit. This is an obstacle to the eradication of the malaria problem. In what follows, we will first look at the tribulations caused by public policy (A) and then at those caused by the drugs that y are sold (B).

A- Tribulations linked to nationals

The notion of public policy beneficiaries refers to the individuals or groups for whom policies are intended[410] . It also refers to the members of a community who are characterised by a similar 'unique competence', and towards whom social policies are directed. Nationals of the same area are characterised by a similar

---

[407] Read AMIEL (P), Ethnomethodologie appliquee: elements de sociologie praxeologique. Presse du Lema, 2010. p. 68.
[408] Ibid. p. 68.
[409] Read COULON (A), L'ethnomethodologie. Paris. PUF (coll. "Que sais-je?", prem. ed. 1987), 1996, p. 57.
[410] See LEVY (J) and WARIN (P), "ressortissants", in BOUSSAGUET (L), JACQUOT (S) and RAVINET (P) (dir.), Dictionnaire des politiques publiques. SciencesPo Les Presses, 5eme edition, 2019, p. 555.

habitus, culture and codes, which is why ethnomethodologists refer to them as 'members'. They often use established protocols or familiar patterns of behaviour to achieve their objectives[411] . In the Centre region, the nationals are not a homogenous reality, as they differ depending on whether they are in rural or urban areas. Moreover, nationals differ according to the district or neighbourhood in which they live. The nationals of the Centre-Cameroon social field have characteristics that cast a shadow over the hardening of the modern fight against malaria. This is why they are a source of tribulation in the social fight against this disease. A distinction must nevertheless be made between people living in rural areas (1) and those living in urban areas (2).

1. Tribulations from people living in rural areas

In the rural areas of Central Cameroon, the inhabitants are generally illiterate, destitute and have a strong tendency to self-medicate, all of which hampers the social fight against malaria. First of all, it should be noted that in the health districts on the outskirts of the Centre, there are many illiterate people, and the language used is vemacular. This state of affairs was not conducive to the awareness-raising phase that preceded the distribution of the LLINs. In the Yoko health district, for example, some health areas are mainly made up of old people, most of whom speak only the vemacular language. The dialogue between the health care providers who came from Yaounde and the beneficiaries living on the spot was not an easy undertaking. A community health worker explained the difficulties he had encountered during the household awareness-raising and LLIN distribution sessions in the following terms: *"In Yoko, the neighbourhood known as 'Yoko' posed a lot of problems for me. There are Pygmy Labas who have difficulty expressing themselves in French. The campaign to raise awareness of how to use mosquito nets was therefore in vain. As a result, the mosquito nets were distributed to most of the households in this neighbourhood without any awareness-raising whatsoever"*[412] . It follows, therefore, that awareness-raising, which is the first line of defence against pandemics, was hampered by the high level of illiteracy among rural populations. Some health workers even claim that in the villages of Donga and Mifoumbe, people speak mainly in the *Baveuk* language and "can't even say hello". The various behaviour change communication (BCC) and home-based malaria management (PECADOM) campaigns are futile in such a social context. When social actors do not share the nationals' 'code', social policy is compromised because 'the code is not the subject of conversations or mundane commentaries between prisoners, it is lived. The code is generally tacit, but at the same time it structures the situation. It becomes part of the language"[413] . For the social fight against malaria to be plausible, the social actors need to be able to become part of

[411] Read HALL (P.A) and TAYLOR ROSEMARY (C.R), "La science politique et les trois neo-institutionnalismes", Op. cit... p. 473.
[412] Interview with Mr Voutsi Pierre on 22 June 2021, who was involved in the various LLIN distribution campaigns as a community health worker in the Yoko district.
[413] Read COULON (A), L'ethnomethodologie. Op. cit... p. 21.

the environment in which they are deployed, through a process of affiliation that will make them members of that village. Clearly, the agents in charge of the social fight against malaria in Yoko were not members of this sub-system, because "a member is someone who has 'mastered the natural language', the social skills of the community in which they live"[414] . Social actors must have "an intimate view of a particular social world", which means sharing a common language with members in order to avoid misinterpretation[415] . It is therefore up to the social actors to imbibe the 'code' of the nationals in order to overcome the constraint represented by the illiteracy of the local populations.

Moreover, according to Cameroon's national malaria control policy, households and communities should play an important role in defining health policies and programmes, and in planning and implementing them[416] . The poverty of people living in rural areas of Central Cameroon is a real obstacle to community participation in the social fight against malaria. Many households do not have enough money to pay for their own treatment. This state of destitution limits the social action of hospitals in the care and treatment of malaria cases, because at district hospital level, there is a hospital policy whereby *"the money collected from 65 patients by a hospital enables it to receive money from the regional delegation for the treatment of 100 patients"[417]* . In other words, the financial resources of a district hospital come mainly from the money collected from the "head" of each patient treated. When patients are indigent, hospital resources dry up. This only serves to undermine the hospital and its staff, as the director of Yoko district hospital told us:

"The huge difficulty we face here in Yoko is that the patients we see in this hospital often don't have the means to pay for their treatment, which is essential for the hospital to function. As a result, we find ourselves having to dig deep into our own pockets to pay for the care of some patients, because the state does not allocate funds directly to health facilities"[418] .

From this statement, it appears that hospitals in rural areas have difficulty in caring for malaria patients because of the destitution of these people. This state of destitution stems from the fact that in the rural sub-systems of the Centre, people live essentially from farming. This activity is much more for self-consumption and less for marketing. As a result, although they are protected from the need for food, these populations are hardly protected from financial need when it comes to paying for health care in hospitals. This state of affairs is a serious blow to hospital coffers and has a financial impact on doctors, who are forced to dip into their own pockets to treat patients who have no financial resources. The social fight against malaria also comes up against another difficulty in rural areas, namely the destitution of the population. This is why we think it would be a good idea for the State not to cancel

---

[414] Ibid. p. 49.

[415] Ibid. p. 49.

[416] Read <u>National Strategic Plan for Malaria Control in Cameroon 2007-2010</u>. Op. cit... p. 22.

[417] Interview with the Director of the Ntui District Hospital on 5 July 2021.

[418] Interview with the Director of Yoko District Hospital on 6 July 2021.

out the initial policy, but to provide rural hospitals with a sum of money which will enable them to deal with cases of real destitution among patients. This is the role of a "reflexive", "supervisory" or "propulsive" State[419] , i.e. a State which masters and takes into account the concrete social realities of its sub-systems.

Moreover, because of their strong attachment to the cultural codes and cognitive scripts of their land, people living in the rural areas of Central Cameroon have a strong propensity for self-medication. In fact, in the health districts on the outskirts of the Centre, despite the hardening of the control method, the large-scale "apolicy" to combat the malaria problem is based on self-medication. In simpler terms, this means managing malaria at home. It should be noted that household management of uncomplicated malaria is a popular practice officially encouraged by the National Malaria Control Programme (NMCP). This practice is also encouraged by poverty and attachment to the socio-cultural register[420] . Among the many stakeholders in Cameroon's health sector, the National Malaria Control Strategic Plan (NMCSP) mentions 'households'. Done Self-medication is a practice that is recognised and encouraged by the regulations and malaria control instruments in Cameroon. However, this practice has become the mainstay of malaria control in the rural districts of central Cameroon, to the extent that it is the only way of combating the disease. However, the NMCP sets the framework within which household treatment is provided, i.e. only in the event of a "simple malaria episode". However, people living in outlying districts flout this prescription by extending self-medication to episodes of severe malaria.

In concrete terms, when we talk about self-medication, we are of course talking about the household management of a disease, but we are also talking about the strong presence of an informal care system and the institutionalisation of parallel health structures. These are the sociological forms of response to malaria that can be observed in the remote areas of the Centre region, and which correspond to the 'production of local systems of public action'[421] . The populations of this part of the region organise themselves and develop a "group sovereignty" which is in permanent transaction with the political sovereignty of the State[422] . This "group sovereignty" enables them to develop their own response strategy to the malaria problem, in line with their cultural codes.

Thus, the treatment of malaria in the periphery of the Centre-Cameroon region is largely the work of households. In this part of the health social field of the Centre, people are strongly bound by 'primarian solidarity' and therefore prefer to deal with their problems within the family unit. Thus, when faced with an episode of malaria,

---

[419] See PAPADOPOULOS (Y), <u>Complexite sociale et politiques publiques</u>. Op. cit... p. 13.

[420] Read MEVA'A ABOMO (D), "Le fardeau de la lutte contre le paludisme urbain au Cameroun : etat des lieux, contraintes et perspectives", RCGT, vol 3 (2), p. 30.

[421] Read LASCOUMES (P) and LE GALES (P), <u>Sociologie de 1 action publique</u>. Armand Colin, 2ᵉ ed. 2012, p. 38.

[422] Read FAURE (A), <u>La question territoriale. Pouvoir locaux. action publique et politique (s')</u>. Political Science, Universite Pierre Mendes-France-Grenoble II, 2002, p. 119.

patients are first treated with a drink prepared in their respective family circles[423] . In the Ntui and Yoko health districts, people are reluctant to be treated in hospital by a doctor, for two reasons. The first is that the medical staff in these districts are constantly criticised by the local population, because the examinations they carry out on the few patients who go there always end up in cases of complicated pathology (hepatitis, AIDS, chlamydia, etc.) which require the payment of huge sums of money. The public believe that this is a strategy for crooked doctors "to make money on the backs of poor patients"[424] . The second reason is that the hospitals in the region's health districts are constantly in a state of indescribable disrepair and insalubrity. Examples include the health facilities in the "Ndimi" health area, such as the Salakounou CS, the Biagnimi CSI and the Ad lucem CS in Ossombe. This list also includes all the health facilities in the "Doume", "Makouri" and "Mankin" health areas in the Yoko health district. In these health centres, the buildings are in a state of disrepair, weeds have taken up residence and in the dry season, the premises are full of dust. This is one of the empirical reasons why people from the Ntui and Yoko districts prefer to be cared for at home, and not without success, rather than go to a health centre for treatment. The other reason lies in the attachment of households in these districts to their cultural register.

Malaria is treated at home using "household pharmacies"[425] . Unwilling to be treated in hospital, households are doing their best to set up home pharmacies. These include medicines such as quinine, nivaquine, ACT, paracetamol and many others designed to relieve the sick within the family structure. These medicines, which are real material resources for households, are imported from Yaounde when a family member travels. In addition to the 'household pharmacies', the populations of the districts in the remote areas of the Centre are strongly attached to 'medicinal plants'.

Indeed, one of the main reasons why people in rural health districts abandon health facilities is their strong attachment to so-called "medicinal" plants. The most popular and sought-after plant is citronella. In the event of a fever, many households do not hesitate to use this plant. The head chief of the village of Ndimi in Mbam-et-Kim explained the virtues of this plant in an interview:

"Citronella is a magic plant that our ancestors used for a long time to fight against disease. I in particular don't go to hospital, even what you call tablets I don't take 9a. I use citronella to deal with my health problems and those of my children. When I have a small fever, I boil it in clean water with a little 'ndjindja' and then I sometimes drink it with a little sugar. Thirty minutes later, I start to sweat, my joints become stronger, my tiredness goes away and I suddenly feel like a young garden"[426] .

Ultimately, the use of medicinal plants and recourse to 'household pharmacies' are

---

[423] See MOULIOM MOUNGBAKOU (I.B), "Concurrence des therapeutiques traditionnelles et biomedicales dans le lutte contre le paludisme a I Extreme-Nord du Cameroun", Op. cit... P. 151.
[424] Interview with a native of Ndimi in the Ntui health district.
[425] A sort of small pharmacy found in people's homes, containing medicines bought in Yaounde, usually kept in a cupboard.
[426] We spoke to Ndimi's top boss, who was kind enough to remain anonymous.

part of the 'social conventions' of individuals living in the villages of the Centre region. When individuals "act according to a social convention, they simultaneously constitute themselves as social actors, i.e. they undertake actions endowed with social significance and reinforce the convention to which they obey"[427] . Hence the high propensity among local people for self-medication, despite the fact that this is excluded from the institutional framework for the fight against severe malaria. This strong tendency to self-medicate is encouraged by the fact that, instead of conventional medicine, the supporters of parallel medicine, who are closer to the people, regularly educate them about the need to keep barks or pharmaceutical products in their pharmacies to prevent any sudden outbreak of fever[428] . What we have seen in the villages of the Centre region is that, at the entrance to every home, there is a good portion of "acitronnelle", which seems to have become the key against any illness that nature has given them. As a result of the use of "household PHARMACIES" and "medicinal HERBS", it can be said that, in place of official preventive and therapeutic measures, people are now resorting to traditional practices or clandestine medicines[429] . All of which hinders the institutionalisation of a modern approach to malaria control in the Centre-Cameroon region.

2. Tribulations from people living in urban areas

In the urban areas of the Centre-Cameroon region, the population is characterised by its density and insalubrity, which complicates the social fight against the malaria problem in urban areas. The density of the population living in urban areas is a serious obstacle to the social fight against malaria. First of all, this state of affairs makes it difficult for urban health centres to cope, as they are overburdened by the number of patients they have to treat. With this in mind, it is worth noting, as Mr MEVA'A ABOMO Dominique does, that THE healthcare system in Cameroon's towns and cities is overloaded. This means that supply is insufficient in relation to demand. It also affects the quality of care"[430] . The real problem lies in the fact that Yaounde's health districts include health areas with more than 50,000 inhabitants! This does not comply with WHO regulations, which are repeated in decree no. 95/040 of 07 March 1995 on the organisation of the Ministry of Public Health in Cameroon. This decree specifies that an urban health area must have a maximum of 12,000 inhabitants[431] . However, the ABIYEM-ASSI 1 and aMelen health areas[432] each have a population of around 60,000. This social reality, which does not comply with WHO regulations, overloads each of the health facilities in these areas and adversely affects the quality of care; all of which encourages the emergence of

---

[427] Read HALL (P.A) and TAYLOR ROSEMARY (C.R), "La science politique et les trois neo-institutionnalismes", Op. cit... p. 484.
[428] See MOULIOM MOUNGBAKOU (I.B), "Concurrence des therapeutiques traditionnelles et biomedicales dans la lutte contre le paludisme a I Extreme-Nord du Cameroun", Op. cit... p. 150.
[429] Ibid. p. 150.
[430] See MEVA'A ABOMO (D), Op. cit... p. 34.
[431] Ibid. p. 34
[432] Melen is the largest health area in the Biyem-Assi district, with a population of 59,059.

"parallel medicine", which is often unsuitable for treating sick people.

Secondly, the population density of the Yaounde districts hinders the deployment of social actors in charge of the social fight against malaria. One of the difficulties encountered by enumerators in 2016 was the large number of households to be visited. There was a glaring discrepancy between the number of households and the number of agents responsible for the door-to-door campaign to distribute coupons. In the vast Biyem-Assi health district, for example, the enumerators had their work cut out for them as they had to deal with a "district-geant" covering two arrondissements of Yaounde (arrondissements III and VI) and made up of almost 400,000 inhabitants, according to the administrative authorities. In such a context, the LLIN distribution campaign was bound to be a failure, especially when the number of household enumerators was limited to 60. The words of a person in charge of enumeration in this district are revealing:

"The household enumeration campaign was not a sinecure in the Biyem-Assi district, as it cost us time, money and energy. There were too many households to cover, and there were only sixty of us. As a result, we were obliged not to count certain households located in rural areas of the district and those in precarious areas"[433] .

It follows from this statement that some households in the Biyem-Assi health district were not enumerated between 18-30 April 2016 and consequently did not receive coupons, which was the prerequisite for receiving the LLINs at the distribution point. This was the case for households living in the "Akok Doe" health area. The density of the population really hampered the distribution sequence of the LLINs in the districts of the urban areas and therefore in the whole of the Centre-Cameroon region. This is why, out of the 3,584,478 LLINs expected and planned for distribution throughout the Centre region, 2,311,626 LLINs were distributed, representing a population coverage rate of 64.4%, below the targets set by the PNLP and MINSANTE! In addition to their density, people living in the urban areas of Centre-Cameroon are also characterised by their insalubrity.

In fact, on observation, most of the population of Yaounde lives in a state of total insalubrity, which does not favour the fight against malaria in the urban areas of the Centre region. The Elig-Edzoa, Biyem-Assi, Essos and Manguier neighbourhoods are obvious examples. Here, household rubbish litters the alleyways, giving off an unhealthy stench. Mirecages and streams of dirty water have taken up residence. Grass is invading the area around the houses. The only cause of all this is the local people, who have turned insalubrity into a class *habit*[434] . In fact, in the Manguier district, local people have been seen dumping rubbish by the roadside at the same time. This suggests that in urban sub-systems, insalubriousness has become a *habitus* understood as a system of durable and transposable dispositions, structures predisposed to function as structuring structures[435] . In other words,

---

[433] Interview with Mr FOUDA Blaise, community health worker in the Biyem-Assi district on 8 July 2021.

[434] Pierre Bourdieu considers that the habitus is nothing other than this immanent law, *the lex insita* inscribed in bodies by identical histories, which is the condition not only for the concertation of practices but also for the practices of concertation. Read BOURDIEU (P), Le sens pratique. Les editions de Minuit, 1980, p. 99.

[435] BOURDIEU (P), Idem. p. 88.

insalubriousness becomes here like an immanent law forming part of the objective living conditions of the populations of Yaounde. In such a context, it becomes very difficult to combat the female anopheles, the vector par excellence of malaria.

In the end, whether in rural or urban areas, the populations of the social health field in Centre-Cameroon are in themselves major tribulations of the dynamics of institutionalisation of the model fight against malaria. In the health districts of rural areas, it is illiteracy, destitution and the strong propensity of the population for 'primarian solidarity' that hinder the social fight against malaria. In urban areas, it is the population density and unsanitary conditions that make it difficult for social players to take action. Nationals are therefore a major constraint to the fight against malaria in the Centre region. In addition to this constraint, the prevailing drug 'policy' in this geographical area is a major obstacle to the process of institutionalising the fight against malaria.

B- The tribulations associated with the "medicamentation" or "pharmaceuticalisation" of anti-malarial drugs

Sjaak Van Der GEEST[436] uses the terms 'medicamentation' and 'pharmaceuticalisation' to describe concepts generally used to describe the growing use of pharmaceutical products to restore and maintain health. Indeed, he writes, "all over the world, people increasingly believe that they need a pill for every health problem they encounter. This craze for drugs is fuelling demand for them"[437] . Medicines therefore become an essential commodity for dealing with public health problems, and as a result they have to be placed on the market in order to guarantee their commercialisation. It is on the basis of this reality that the drug becomes an object of "commodification"[438] , i.e. an economic value is now assigned to it because it is now part of the circuit of buying and selling.

One of the most common ways of combating malaria is to provide the population with anti-malarial drugs, which are medicines designed to halt the development of plasmodium in the human body. This means that each State should adopt a drug "policy" to protect its citizens from major health problems. From this point of view, medicines appear to be "health goods" which must be made available to the population. Mr PODA BOUMANAI Angelain said that "in reality, medicine is a 'health good'. Consequently, if health is a public good, the drug that is the vehicle for health is an 'intermediary' public good that leads to health, which is the ultimate public good"[439] . We cannot talk about public health issues without mentioning the crucial role of medicines in this process. But one fundamental issue remains: access to quality medicines for all. Christophe FOE NDI takes it on board when he writes

---

[436] Read GEEST (SVD) "Les medicaments sur un marché camerounais. Reconsideration de la commodification et de la pharmaceuticalisation de la sante", In <u>Revue Internationale francophone d'anthropologie de la sante</u>. Association Amades, Varia, 2017, p. 3.

[437] Ibid. P. 2.

[438] This expression means that an economic value is assigned to a thing that was originally conceived outside the circuit of buying and selling. This is the case with health, which is guaranteed by day-to-day commercial relations through the purchase of medicines in places reserved for commerce. For more details on this expression see GEEST (SVD), Idem. p. 3.

[439] Quoted by FOE NDI (C), <u>La mise en auvre du droit a la sante au Cameroun.</u> Op. cit... p. 214.

that "access to medicines for all is seen as a fundamental human right because it is an essential complement and component of the right to health"[440] . He goes on to write "However, despite its fundamental nature for the realisation of the right to health, access to quality medicines remains an objective that is virtually unattainable in some countries"[441] . Cameroon is no exception, as access to quality medicines is an acute problem in the Centre region.

The Centre-Cameroon sub-system is characterised by a vast market for the marketing of medicines in general and antimalarial drugs in particular. The problem here is not access to medicines, but access to quality medicines. Finding a quality antimalarial drug is not within everyone's reach in the region, because 'pharmaceuticalization' is most often the work of an informal market and counterfeiting, or even smuggling. This reality is not confined to the Central region, but applies to the whole of Africa, as Edrich Nathanael TSOTSA tells us   : "the

he distribution of medicines and other pharmaceutical products through an informal network is a reality that can be observed in the day-to-day environment of the global trading conditions for goods and services in Africa"[442] . As medicines are 'health goods', their poor quality constitutes a serious breach in the social fight against disease. In the social health field of Centre-Cameroon, the drug is even becoming a constraint on efforts to combat malaria, insofar as it suffers from a glaring lack of quality, because its circulation is the result of informal sales (1) and counterfeiting (2).

1. The tribulations associated with the informal marketing of anti-malarial drugs

In the Centre-Cameroon region, there is a strong informal market in anti-malarial drugs, which people use to provide their own healthcare. It is clearly gaining ground at the expense of conventional pharmacies and the public health system.

Observation shows that the informal private drug sector grows as you move from the centre of Yaounde to the outskirts. It exists where, and because, public services are not achieving their objectives. Informal medicines are more available and more accessible to people than those from the formal private and public sectors, for three reasons. Firstly, the medicines sold by vendors are the most affordable. With them, customers can buy the quantity they need to practise self-medication at a given time, unlike conventional pharmacists who only sell medicines in their original packaging, which is too expensive for some customers. Secondly, informal drug sellers are geographically more accessible than conventional pharmacies, as they can be found within a few kilometres of the patient's home, unlike pharmacies or dispensaries, which are often located 50 kilometres from the patient's home. Finally, most vendors are available day and night. Their shops close when everyone

---

[440] FOE NDI (C), Idem. p. 214.
[441] Ibid. pp. 214-215.
[442] See TSOTSA (E-N), Op. cit... p. 297.

is in bed, and after closing it is still possible to buy medicines, as some vendors have their homes close to their shops. This flexibility contrasts sharply with the strict working hours of formal services. For these three reasons, the informal antimalarial sector is taking precedence over the formal sector. As a result, most malaria patients are treated with antimalarials from an informal and unconventional circuit.

As a result, it has been observed that the social health field in Central Cameroon is one in which the informal private drug sector has the wind in its sails. This is encouraged by the "popular belief" that drugs are the "miracle solution" to health problems. In Yaounde, drugs have even become a precious commodity, essential to a truly healthy life, so much so that buying them is seen as the most rational thing to do. NICHTER underlines the 'false consciousness' implicit in this type of behaviour when he states that "an illusory sense of health security is fostered by the exaggerated claims of pharmaceuticals..."[443] . He thus introduces the term "pharmaceuticalisation", a process he sees as a form of "fetishisation": the attribution of power to medicines beyond their active ingredients[451] . Medicines are seen as the 'miracle solution' to the process of achieving good health, and this belief encourages the widespread purchase of medicines and the development of an informal market for them.

In the geographical area of the Centre, informal sector drug sellers occupy the space for marketing anti-malarial drugs, which they make available to the local population. These vendors constitute heterogeneous groups, which favours their multiplication in the region. Most of them are ordinary merchants who sell everyday products, including antimalarials, in shops or kiosks. In the city of Yaounde, there are about 800 such shops and kiosks where you can buy at least one or two types of antimalarial drugs, the most common being Quinacrine, Nivaquine, Quinine, Chloroquine, Sulfadoxine-Pyrimethamine, Amodiaquine and Artemisimin- base combination therapy. A second category of informal medicine sellers is made up of 'colporteurs' who travel from village to village during the cocoa harvest season, when villagers have a little money. These 'colporteurs' sell several kinds of goods as well as medicines. A third category is made up of market vendors who sell medicines among other products. They can be found in large numbers in Yaounde in the "central" markets of "Mfoundi" and "Elig-Edzoa". A fourth category of vendors are those who specialise in the sale of medicines and whose range of products is much wider than that of the vendors mentioned above.

Informal vendors occupy the space where anti-malarial drugs are marketed and are located close to the population, so they are in greater demand than traditional pharmacists. Although the medicines they make available to the population are of dubious quality and efficacy, households prefer to drink from their sources rather than go to a pharmacy or hospital, which are often very far from where they live, and when they know that the medicines available there are expensive. This reality

---

[443] Quoted by GEEST (SVD), Op. cit... p. 9.

is a constraint on the social fight against malaria, given that the informal marketing of anti-malarial drugs is carried out by untrained vendors, and the medicines sold are of unknown origin. As a result, rather than looking after their own health, people are putting themselves in a position where they can either contract another disease or not treat the malaria problem at all, because the anti-malarial drugs they are taking are of dubious quality. Despite the efforts made by the public authorities to combat the informal sale of medicines[444] , it should be noted that this practice persists, and that this persistence is the immediate result of the poverty of the population, who see the informal sector as an "abon marché" for buying cheap anti-malarial drugs. Sjaak Van Der GEEST goes further, pointing out that it was the failure of the public health system that encouraged the emergence of the informal private drug sector when he writes: "these private services - both formal and informal - were living proof of the dysfunction of the public health system"[445] .

In addition, the informal marketing of anti-malarial drugs encourages the promotion of so-called "parallel pharmacies". Also known as "clandestine amedecines"[446] , "APHARMACIES paralleles" refer to cheap therapies dispensed by backstreet pharmacists or in homes by trained nurses not recruited by the state. These pharmacies exist in large numbers in the health districts on the outskirts of the Centre region and are not subject to any prohibition by the public authorities. They are practically accepted as the normal way of treating illnesses, alongside conventional pharmacies. As a result, they are practically institutionalised by the State. The distribution of medicines and other pharmaceutical products through an informal trading network is a reality that is part of the daily environment of the global conditions of trade in goods and services in Africa[447] .

In the Ntui health district, there is a strong presence of "parallel PHARMACIES", where medicines are sold to the public at very low prices. These apharmacies are replacing conventional pharmacies, where the cost of medicines is considered too high. A well-known clandestine supplier known as "Bamenda Boy" has a "parallel pharmacy" right in the heart of the town of Ntui at a place called "Centre". He is a doctor who has undergone training but has not been recruited by the state. He has a large stock of medicines, syringes and infusions, enabling him to treat people suffering from malaria at home. Another clandestine provider, based in the Haoussa district, treats malaria patients in his home at very reasonable prices. All in all, there were nine "parallel pharmacies" in the "Ntui" health area, compared with two conventional pharmacies approved by the state[448] . Here, clandestine medicine seems to have become a norm, a substitute, a palliative for medical deserts. In this respect, they are an obstacle to effective and efficient social control

---

[444] The latest effort comes from Minister Manaouda Malachie, who issued a statement on 10 July 2019 banning the sale of medicines from the informal market.

[445] See GEEST (SVD), op. cit... p. 7.

[446] See MOULIOM MOUNGBAKOU (LB), Idem. p. 150.

[447] See TSOTSA (E-N), Idem. p. 297.

[448] These are the pharmacies known as "sainte Therese de Ntui" and "pro-pharmacie la perseverance", located in the heart of the town of Ntui at a place called "centre".

of the malaria problem.

In the Yoko health district, "parallel pharmacies" are also in full swing. In this social health field, it is the presence of "medical deserts" that encourages the emergence of these pharmacies. A notorious clandestine doctor known as "patson" provides the public with medicines of all kinds to relieve malaria sufferers. There are four paralegal pharmacies in this sub-system, compared with two conventional state-approved pharmacies. In the health districts on the outskirts of Centre-Cameroon, clandestine medicine or parallel pharmacies are an integral part of the "measures" to combat malaria. Although "clandestine" or "parallel", these pharmacies are now considered to be the normal and institutional framework for combating the malaria problem, given the number of people who go there when they are ill. They are thus replacing conventional pharmacies, which are virtually non-existent in this geographical area. Because of their dubious effectiveness, "parallel pharmacies" are a major constraint on the drive to institutionalise the fight against malaria. The street sale of medicines or "street medicine" is also included in the category of so-called "parallel" malaria control structures. In other words, in addition to the problem of the informal marketing of anti-malarial drugs, there is the problem of counterfeiting.

2. The tribulations associated with the widespread sale of counterfeit medicines

In addition to the informal sale of anti-malarial drugs, which has the wind in its sails in the Centre region, there is also the phenomenon of counterfeiting, which is having a major impact on anti-malarial drugs. This is a blatant distortion of the social fight against malaria. Monique MAS describes the calamitous situation of drug counterfeiting in Cameroon and throughout Africa in the following terms:

"Some counterfeit medicines are manufactured and packaged in tins or bottles under fanciful names, or under names that have actually been listed by the authorised pharmacological services. In other words, there are clandestine production lines that copy the packaging of pharmaceutical products, disregarding the molecular content of the products. Local corrupt individuals are responsible for injecting them into poorly supervised local distribution circuits. Fake antimalarials in Cameroon, fake antibiotics in the Democratic Republic of Congo, these counterfeit products are not only sold over the counter in marabout's shops. They also find their way into pharmacies and hospitals, their ineffectiveness contributing to the spread of the diseases they are <.-.   457 censees traiter " .

The people living in the social and healthcare sector of Central Cameroon are not immune to the dangers of counterfeit medicines, as they live in an area where the phenomenon is in full swing. In fact, there is a "small" drug market here, made up mainly of "street" sellers. This "small" drugs market is a direct result of poverty and the growth of the informal sector. These street vendors of counterfeit medicines, who operate in the open air in Yaounde in full view of the public, obtain their supplies in Douala, the country's economic capital; those in Douala obtain their supplies in Bamenda and the surrounding area; those in Bamenda find supply relays from the Nigerian market [449][450] . In other words, the fake antimalarial drugs circulating on the market in the Central region come from Nigeria. Unemployed young people see these counterfeit drugs as a way of making money and, above all, creating a job. In Yaounde and the surrounding area, sellers of "counterfeit" medicines have open-air dispensaries in market squares, at hospital entrances, or even sell medicines on the move in road and rail stations, and even on trains and in hospitals[451] . Many people are thus exposed to these counterfeit medicines, which are harmful to health and can contribute to the spread of the diseases they are supposed to treat. This was the experience of Mrs Ombolo Albertine, who lives in the Mimboman district of Yaounde. After consuming a medicine bought from a street vendor, she found herself in a critical condition. Elie tells us all about what she endured in an interview:

"One night I felt very ill. I had a fever that kept rising, violent headaches and nausea. I knew straight away that it was malaria. I always experience the same symptoms when I want to get sick. Usually, when I have symptoms like that, I take quinine mixed with paracetamol or simply efferalgans. But that night, I bought my supplies from a young street vendor who had a large quantity of these tablets in a basket. The reason for this was that I was all alone at home and I was feeling a bit under the weather. I managed to buy two blister packs of quinine and two blister packs of paracetamol from this vendor for 400 francs. I drank the tablets before going to sleep. Instead of calming me down, 9a actually made me sicker. The headache, fever and nausea persisted to such an extent that I didn't close l'mil all night. The next morning, I went to the hospital"[452] .

This statement clearly reflects the danger posed by counterfeit medicines being sold in the open air in the

Centre. Instead of producing a positive result, the ineffectiveness of these medicines actually contributes to the spread of the disease they are supposed to treat. As the number of street vendors increases and poverty becomes more widespread, many households find themselves unwittingly exposed to the mercy of counterfeit antimalarials. For Edrich Nathanael TSOTSA, "it is the poor who suffer the effects of counterfeit drugs and dubious supplies. They represent communities where the illiteracy rate is very high, so they are not very receptive to awareness-raising about the existence of counterfeit medicines in pharmaceutical distribution networks"[453] . Poor people in poor health have no choice but to buy cheaper medicines of dubious and illicit origin[454] . In such a context, it becomes difficult to carry out a moderate and effective fight against the malaria problem.

Given the scale of the intensive trade in medicines between Nigeria and its neighbours, a large parallel distribution network has developed around informal networks of medicines and pharmaceutical products that now supply the "small"

---

[457] Monique MAS, "Sante, la corruption qui tue", RFI, Paris, 11 May 2006, quoted by TSOTSA (E.N), Op. cit... p. 303.
[450] See TSOTSA (E-N), Idem. p. 299.
[451] Ibid. pp. 298-299.
[452] Interview with Mrs Ombolo Albertine, resident of Yaounde in the Minboman district on 14 July 2021.
[453] See TSOTSA (E-N), Op. cit... pp. 301-302.
[454] Ibid. p. 302.

Cameroonian market[455] . This has given rise to the sale of street medicines, which has found fertile ground in the rural areas of the Centre region. This "small" market is made up of "street" drug sellers. These people run open-air dispensaries in market squares, at hospital entrances, or even sell medicines on the move in road and rail stations, and even on trains and in hospitals[456] . In fact, street vendors in the Centre region obtain their supplies in Douala; those in Douala obtain their supplies in Bamenda and the surrounding area; those in Bamenda find supply relays from the Nigerian market. Some street vendors are supplied directly by the official drug stock management networks: doctors, laboratory assistants, medical representatives, etc., who get rich from these illicit networks.[457]

In the Centre-Cameroon region, districts in rural areas are the "places" par excellence for marketing, buying and distributing street medicines. These medicines, most of which are counterfeit, are displayed in front of roadside stations and markets. They are sold at low prices by people who have no qualifications in the medical or pharmaceutical field. Because they are poor, many people buy their supplies from these "street vendors" in the event of an episode of simple or severe malaria. These vendors have drugs as important as chloroquine, mefloquine, halofantrine, quinine, pyrimethamine, proguanil, sulphadoxine and artemisinin, which are among the most widely available essential antimalarial drugs.

In the Ntui and Yoko health districts, street drug sellers are the undisputed masters of anti-malarial drug distribution. In the "Ngoro", "Ndimi", "Ndjame", "Nguila" and "Talba" health areas, street drugs are sold like bread. When a sick person is unable to find what they need at home or at a herbalist's, their first instinct is to go to the "cheap" street drug sellers. Poor people in poor health have no choice but to buy the cheaper medicines, of dubious origin and illicit[458] . In other words, it is the poverty of the local population that encourages the emergence and spread of street medicines. The same is true of the "Mankim", "Nditam" and "Ndjole" health areas in the Yoko district.

It should be noted, however, that the sale of street medicines falls into the category of "illicit" trade and is punishable under the law of 10 August 1990[459] . In a communiqué dated 10 July 2019, MINSANTE undertook to put an end to this type of trade in Cameroon by criss-crossing the country's towns and cities with a view to seizing and destroying all medicines sold in the street. In the Centre-Cameroon region, the operation was limited to the city of Yaounde, where street drugs are still sold.

Illicit medicines continue to be sold in rural areas such as the Ntui and Yoko districts. It could not be any other way, given that the majority of the population in

---

[455] Ibid. p. 298.
[456] Ibid. pp. 298-299.
[457] Ibid. p. 299.
[458] Ibid. p. 302.
[459] See law №90-035 of 10 August 1990 on the practice and organisation of the profession of pharmacist, article 53 of which states that "any offence, display or distribution of medicines is prohibited on the public highway, in fairs and markets to any person, even those holding a pharmacist's diploma".

rural areas live below the poverty line, and that conventional pharmacies y are rare, and even when they do exist the prices y are extremely high. The reality of the territory has indeed prevailed over the logic of the state, to such an extent that the sale of street medicines has become the norm in the rural areas of the region, and is now part of the normal framework for the fight against malaria in these sub-systems, especially as a large proportion of the population claim to find relief through these medicines obtained from parallel pharmacies.

What is even more serious is the fact that some street vendors are directly supplied by the official drug stock management networks: doctors, laboratory assistants, medical representatives, etc., who make their money from these illicit networks. At the same time, these vendors are also able to supply public and private facilities, either by reselling medicines obtained from their official intermediaries, or by simply delivering those brought in from Nigerian networks at low prices via the "Bamenda Boys"[460] . The sale of counterfeit drugs is not confined to the street, but also affects conventional and public drug distribution structures. Faced with a difficult financial situation, some doctors and pharmacists obtain anti-malarial drugs from the informal market at a low price and sell them to patients at the conventional price. As Jean-Pierre Olivier de Sardan shows, the trend towards corruption that is undermining the health sector is now affecting all sectors: nurses, doctors, administrators and association staff. All of them sometimes misappropriate funds, equipment or even medicines for personal gain[461] . The desire of healthcare staff to "make ends meet" sometimes leads them to "informalise" their activities. As a result, there are problems with the quality of medicines available in health centres: some medicines are perishable;

Others are sometimes kept in unsuitable conditions; still others are recommended by the doctor as an alternative to the medication available[462] .

In such a context, it becomes difficult for patients to avoid counterfeit medicines, whatever their financial means. The fight against malaria is thus dealt a severe blow, because instead of being a social instrument for treating malaria, the drug is becoming an object of commodification, pharmaceuticalization and enrichment in the social health field of Central Cameroon. Here, it matters little whether the drug is of good quality or not; what matters is how it is marketed. According to the World Health Assembly, the unreliable supply and quality of medicines is one of the problems hindering people's access to quality care and safety in developing countries[463] . Drug counterfeiting is therefore a real constraint on the drive to institutionalise modern malaria control in the Centre region.

In conclusion, a number of tribulations punctuate the process of institutionalising the fight against malaria in the Centre region. Whether at institutional or social level, the actors in charge of implementing the intensification of the model malaria

---

[460] See TSOTSA (E-N), Op. cit... p. 299.
[461] Ibid. p. 291.
[462] Ibid. p. 303.
[463] See FOE NDI (C), op. cit... p. 215.

control are encountering difficulties. In short, in the social and health field of Central Cameroon, the institutionalisation of the fight against malaria is not taking place on a 'smooth' terrain, but on a terrain strewn with pitfalls which make the power of the institutional and social actors involved in the construction and implementation of modem malaria control actions derisory. Some tribulations are directly linked to health institutions and make the fight against malaria institutionally complex: this is 'institutional complexity'. Others, on the other hand, are linked to the geographical space in which social actors are deployed and make public action against malaria socially complex: this is 'territorial and social complexity'.

The tribulations linked to health institutions are seriously hampering the institutional fight against malaria. This is reflected first and foremost in the entanglement of malaria control structures. This entanglement can be vertical or horizontal. The explicit corollary of this is the 'overlap' between 'central' and 'peripheral' health institutions and the fragmented political-administrative arrangement between public health services. Secondly, the tribulations associated with health institutions also result from the lack of positive coordination between health facilities. The immediate corollary of this is the lack of permanent cooperation and the "complexity of joint action" between hospitals. These institutional tribulations are a serious obstacle to the implementation of decisions taken by central public players (the President of the Republic and MINSANTE) within health institutions. As Bruno PALIER and Yves SUREL state, "it is undoubtedly in the field of social policy that the weight of institutions has been analysed the most (no doubt because it is an area of public action that is particularly 'saturated' with institutions), insofar as these institutions, in the broadest sense, can influence the nature of the problems encountered, the resources and repertoires mobilised by the actors involved, and the diagnoses and solutions adopted"[464] .

In addition to the institutional tribulations, other so-called socio-territorial tribulations influence the activity of those implementing modem malaria control actions in the social health field of Central Cameroon. These tribulations hinder the action of social actors and bias the social fight against malaria. These tribulations are directly linked to the social space in which the actors are deployed, and are referred to as 'territorial and social complexity'. Here, the tribulations result from the territory, the nationals of this territory and the 'medicamentation' that y prevails. First of all, the territory in which the social actors are deployed represents a serious departure from the sequences of MU distribution and household enumeration, whether in the rural sub-systems or in the Yaounde sub-systems. In the rural health districts, the hostility, the extreme isolation of the area and the presence of medical deserts are all constraints on the dynamics of institutionalising the fight against malaria. In the health districts of Yaounde, it is the vastness of the territory and the

---

[464] See PALIER (B) and SUREL (Y), "Les "trois I" et l'analyse de 1 Etat en action", Op. cit... p. 13.

predominance of 'ecologically favourable areas' that hamper the actions of social actors such as community health and enumeration agents.

Secondly, the people for whom public malaria policies are intended and the anti-malarial medication which prevails constitute serious tribulations to the hardening of the social fight against the malaria problem. In rural areas, illiteracy and poverty make the social fight against malaria more difficult. In urban areas, it is the population density and unsanitary conditions that make the fight against malaria more complex. The prescription of antimalarial drugs for these nationals also represents a serious departure from the traditional fight against malaria, as these drugs are most often sold on the informal market and are not immune to the risks of counterfeiting. In such a context, rather than being an instrument for the treatment of malaria, the drug becomes a potential danger to the health of the population and makes the process of institutionalising the fight against malaria in the geographical area of Centre-Cameroon derisory. As a result, this dynamic is producing ambivalent effects on the social field.

Chapter 4

The relativity of the effects of the institutionalisation of modern malaria control and the need for stronger action

Although it has been through a number of ups and downs, the process of institutionalising the fight against malaria is having an impact (albeit an ambivalent one) on the social environment in Central Cameroon. A study of the effects of a policy's institutionalisation process seeks to establish whether the intended target groups have actually changed their behaviour (what impacts?) and whether, subsequently, the situation of the final beneficiaries, initially considered problematic, has actually improved (what outcomes?)[473] . In other words, talking about the effects of institutionalising a policy means studying its impacts and *outcomes in* the social field (effects=impacts+outcomes). Impacts, which are the effects of a policy that can be observed among the target groups, are defined as "all the changes in the behaviour of the target groups, whether intended or not, that are directly attributable to the entry into force of the P.P.A., the A.P.A., the action plans and the acts of implementation (outputs) that put them into practice"[474] . The aim here is to study the direct effects of the institutionalisation of a policy on the target groups defined as responsible for the public problem to be solved[475] .

The *outcomes, which are* the effects of a policy that can be observed among the end beneficiaries, are defined as "all the effects, in terms of the public problem to be solved, that are attributable to the change in behaviour of the target groups (impacts) themselves induced by the acts of implementation (outputs)"[476] . The aim here is to study the direct effects of the institutionalisation of a policy on the *end* beneficiaries, defined as the main victims of the problem to be solved. The *target groups* and the *end beneficiaries* are therefore the actors on which the analysis of the effects of the dynamics of the institutionalisation of a policy is based.

To deal with the question of the 'effects' of the institutionalisation of a policy is still to analyse the direct result of that policy on the populations, some of whom, by their behaviour, are responsible for the disastrous situation that prevails, while others are the immediate victims of that situation.

In the geographical area of Centre-Cameroon, the institutionalisation of the fight

---

[473] Read KNOEPFEL (P), LARRUE (C), VARONE (F) and SAVARD (J-F), <u>Analyse et pilotage des politiques publiques</u>. Op. cit... P. 301.
[474] Ibid. p. 302.
[475] Ibid. p. 302.
[476] Ibid. p. 307.

against malaria began during the colonial period in 1919, and was reinforced in 2002 with the adoption of the *"roll back malaria"* programme by the Cameroonian public authorities. Since malaria has become a "public problem", solving it has required action from public and private players, as well as civil society. This led to the institutionalisation of "multi-factorial public action on malaria", a concrete expression of the toughening of the modern fight against the disease. As this intensification did not take place in a 'vacuum', a number of institutional and social tribulations hampered the action of the socio-institutional players in charge of this process, as shown in the previous chapter. As a result, the process of institutionalising the fight against malaria has had contrasting effects on the social field. In other words, despite the fact that a whole range of actors have been involved in modern malaria control measures, it has to be said that the disease still persists in the Central Cameroon region, and traditional malaria control methods have not completely disappeared. Hence the ambivalence of the effects of the dynamics of the institutionalization of the traditional fight against malaria on the social field **(Section 1).** In spite of everything, this form of 'model' malaria control can still be improved **(Section 2).**

SECTION I

(... the ambivalent effects of the institutionalisation of modern malaria control

Analysing the effects of the dynamics of institutionalisation of the modern fight against malaria also means analysing its effects on nationals. The notion of nationals refers to the individuals or groups for whom policies are intended[477] . They are also the 'members' of a community towards whom public policies are directed and are expected to produce effects. There are two main stakeholders in public policy: *target groups* and *end* beneficiaries. To these can be added the private actors indirectly affected *by* policy. The institutionalisation of modern malaria control in the Centre-Cameroon region has an impact on both target groups and end beneficiaries. The target groups are defined as all the people who, through their behaviour and attitudes, cause the problem to be solved. The final beneficiaries are those who are the immediate victims.

Although fraught with tribulations, the institutionalisation of the modern fight against malaria in the social health field of Centre-Cameroon has ultimately had an effect on the nationals. It should be noted that some of these nationals are responsible for the large-scale spread of malaria through their behaviour, which tends to promote insalubrity. In fact, before the *"roll back malaria" programme was* adopted and implemented in the early 2000s, certain individuals displayed attitudes of ecological incivism and disregard for basic hygiene measures. These individuals were not only putting their families at risk, but also their immediate surroundings. They were not only a danger to themselves, but also to third parties

---

[477] See LEVY (J) and WARIN (P), "ressortissants", Op. cit... p. 555.

who are "collateral victims" of their actions (*negatively affected third parties*). These individuals constitute *target groups*, and tougher malaria control measures are expected to have an impact on them if the disease is to be reduced.

It should also be noted that some of these nationals are immediate victims of malaria. These are the people who suffer the horrors of the disease on a daily basis, putting their families and loved ones at risk. In fact, malaria sufferers are not malaria sufferers alone, as their illness indirectly affects third parties who are the "collateral victims" of their suffering. As a result, stepping up the fight against malaria, which is designed to limit the number of cases of the disease, will indirectly benefit third parties *(positively affected third parties)*.

To speak of the ambivalence of the effects of the institutionalization of malaria control is to admit that, far from totally containing the threat of malaria, this dynamic produces contrasting effects on the social field. In other words, in spite of the many effects of the development of malaria control on the population, it should be noted that the problem-problem still persists in the Centre-Cameroon region, and traditional methods of malaria control have not completely disappeared. In the remainder of this paper, we will show that, despite the many setbacks that have beset the process of institutionalising traditional malaria control, it has nonetheless led to changes in behaviour and an improvement in the socio-economic conditions of the population **(paragraph 1).** This positive result should not, however, lead us to believe that the malaria threat has completely disappeared **(paragraph 2).**

Paragraph 1: Relative change in behaviour and (improvement in people's socio-economic conditions

In the Centre-Cameroon region, the major effects of the process of institutionalising malaria control have been directly felt by the local population. Their behaviour towards malaria has completely changed, and their socio-economic situation has clearly improved. In fact, the tougher approach to the fight against malaria first had an effect on the *target* groups. It should be remembered that these are not only the people responsible for the collective problem to be solved, but also those through whom the resolution of the problem or its mitigation will be made possible.

From 2002 onwards, a range of actors stepped up communication about malaria, mainly to bring about a change in the behaviour of target groups considered to be responsible for the spread of the disease in the Centre-Cameroon region. It would appear, therefore, that this intense communication effort was not in vain, as y in the end there was persuasion, i.e. an endogenous change in the preferences of an interlocutor or an audience[478] . In other words, the people of the Centre region finally changed their attitude towards malaria by highlighting ways of preventing the disease and respecting them as best they could. If the target groups have been able to change their behaviour, the situation of the final beneficiaries will be significantly improved.

---

[478] Read GERSTLE (J) and PIAR (C), La communication politique. Op. cit... p. 97.

The intensification of the fight against malaria from 2002 onwards then had an impact on the end beneficiaries. *End beneficiaries* are individuals (natural or legal) and their associations who are directly affected by the collective problem, and who suffer its negative effects. These are people who expect their socio-economic situation to improve as a result of the problem being resolved. Their situation actually depends on that of the target groups. Now, if the people whose actions were deemed responsible for the spread of malaria have changed their behaviour, those who suffered from the disease will see their social and economic situation improve. Clearly, the effects of the dynamics of the institutionalization of modern malaria control are materialized in the geographical area of Centre-Cameroon by a change in the behaviour of local people (A) and by an improvement in their socio-economic conditions (B).

A- Relative change in the behaviour of nationals

The dynamics of the institutionalization of the model malaria control initiated in 1919 and reinforced in 2002, had a strong impact on the nationals of the Centre-Cameroon region because they finally modified their behaviour at the origin of the spread of the disease. The aim here is to analyse the impact of the advent of modern malaria control on the target groups considered to be at the origin of the spread of malaria in the region. Most of these people live in poor neighbourhoods and enclaves such as Elig-Dzoa, Manguier, Nkolmesseng, Etoa-Meki, the lowlands of Mballa II and in rural areas of the region. These target groups have been the focus of the tougher anti-malaria campaign since 2002, because the public authorities want them to change the behaviour that causes the disease to spread. Programmes such as *"Nightwatch", the "behaviour change communication campaign"* (CCC), *"K.O paludisme", "tous unis contre le paludisme"* and the many awareness-raising campaigns run by NGOs and socio-political players all aimed to have a significant impact on the behaviour of target groups. Impacts are defined as all changes in the behaviour of these groups that are directly attributable to the implementation of public policies. Measuring the impact of the tougher malaria control measures on the target groups in Centre-Cameroon means considering the following questions: have the target groups changed their unhealthy behaviour? What behaviour do they now adopt? How has their behaviour changed?

In the geographical area of Centre-Cameroon, the process of institutionalising the modern fight against malaria has indeed had an impact on the behaviour of the target groups, despite the many setbacks faced by the implementers. The change in the behaviour of target groups can be seen here in the increased use of hygiene measures **(1)** and prevention measures **(2)**.

1. Intensification of hygiene measures

If we compare the way people practised hygiene measures before and after the *"roll back malaria" programme was implemented*, we can see a marked change in their behaviour. Hygiene measures are being stepped up and scrupulously observed, even if pockets of resistance remain. Before 2002, the people of Central Cameroon were living in a highly unsanitary state; today, things are much better. There is a natural tendency to want to respect basic hygiene measures.

Prior to 2002, almost 74% of households in the Biyem-Assi district lived near an unimproved watercourse (stream, river, lake, marsh). In 2020, around 30% of households y living[479] . The number of households living beside an undeveloped watercourse has fallen dramatically. The residents of this district have become fully involved in treating the stagnant water and collecting household waste near the water, following decisions by the local public authorities to organise "hygiene and sanitation" campaigns during the 20142015 period. They were assisted by the municipal authorities in clearing and levelling the marshy areas, although there is still work to be done, as almost 30% of households still live next to unimproved watercourses. We can see, therefore, that it is the change in people's behaviour that has helped to reduce the number of stagnant watercourses.

In the same vein, it should be noted that the vast majority of people in the Centre-Cameroon region have moved on from using "unimproved toilets" to using "improved toilets". The former are those shared with other households and where human excreta are not effectively separated from all human contact. The second are those that effectively separate human excreta from all human contact (flush toilets, improved and anti-airing cesspits, etc.). Before the 2000s, almost 70% of households in the region used "unimproved toilets", a rate of 95% in rural areas and 25% in urban areas[480] . By 2020, however, 42% of households (65.3% in rural areas and 15.1% in urban areas) were using unimproved toilets[481] . This shows that the number of people using unfinished toilets has fallen considerably, depending on whether the *"roll back malaria" programme came into force* or not.

Today, in the Yaounde sub-system, almost all households use "improved toilets". Only 15% of the population continue to use unimproved toilets. This section of the population is found in neighbourhoods such as Etoa-Meki, Elig-Edzoa, Nkolmesseng and Mballa II. It's in rural areas that people have to deal with "unfinished toilets" on a daily basis. This is due to the prevailing poverty. Here, 65% of toilets are shared, and sometimes ten households share the same toilet, not counting the number of people who use them occasionally and sporadically, since these toilets are located in the open air, often close to back alleys. In rural areas, the large number of "unimproved toilets" is directly linked to the poverty of the people, who are unable to afford proper toilets. However, throughout the region, people have clearly improved their behaviour with regard to the maintenance and fitting out of latrines, as more than half of them use fitted toilets, i.e. a percentage of 58%,

---

[479] Figures obtained from a sample of twenty (20) households in the Biyem-Assi district. The households chosen belong to the category known as "family homes", which had already been established before the year 2000. Following an interview with the heads of these households on 17 July 2021, it emerged that of the 20 households, 13 claim to have lived near unmaintained watercourses before the 2000s and no longer live there today. Seven (07) households claim to have lived there since the 1990s.

[480] Figures obtained from a sample of forty (40) households, including 20 in rural areas (Ntui and Yoko) and 20 in urban areas (Biyem-Assi and Ndjoungolo).

[481] See the document entitled <u>Enquete post campagne sur l utilisation des MILDA 2016/2017. Rapport final</u>. Op. cit... p. 35.

compared with 42% of the population who use un-fitted toilets, the majority of whom live in rural areas. From this point of view, it can be said that the "hygiene and sanitation" and behaviour change communication (BCC) campaigns had a significant impact on the behaviour of the target groups.

What's more, unlike the situation before the 2000s, drinking water is now supplied from "improved sources" such as water taps, treated boreholes and shop-bought mineral water. There is a clear trend for the population to abandon drinking water from 'unimproved sources' (well water, river water, lake water and borehole water), as a study carried out in 2017 showed that 79.2% of households in Cameroon consume water from an 'unimproved source'.

"[482] . By deduction, we can say that in the Centre region, this rate is around 80%. This shows that people are becoming increasingly aware of the need to drink clean water to avoid illnesses such as malaria, even though a large proportion of those living in rural areas continue to obtain their drinking water from unimproved sources.

Since the implementation of the *"roll back malaria"* programme in 2002, people living in the social and health care sector of Centre-Cameroon have been increasingly promoting cleanliness in their living environment. Cleaning up around their homes is becoming a recurrent part of people's daily lives. It involves clearing grass and collecting household rubbish. What's more, there is even a day devoted to cleaning up public spaces, known as "Clean Thursday"[483] . The aim of all these acts of cleanliness is to protect people from unsanitary conditions and prevent the spread of mosquitoes.

In short, we can see that with the implementation of programmes such as *"K.O malaria"* and *"behaviour change communication"* (BCC), people in the Centre region have effectively changed their behaviour. In other words, the tougher approach to the fight against malaria from 2002 onwards has had a significant impact on the target groups, who now pay close attention to basic hygiene measures. The impact is even more significant when we see that people are actually taking malaria prevention measures on board.

2. Stepping up prevention measures

Malaria prevention focuses on two main areas: vector control and chemoprevention in pregnant women[484] . Vector control is based on the use of ITNs/MILDAs, while chemoprevention is based on the use of IPTs. In public health, prevention is always better than response or treatment. This is why programmes such as *"nightwatch"*, *"K.O paludisme"*, *"tous unis contre le paludisme"* and *"communication pour le changement de comportement"* (communication for behavioural change), accompanied by the many awareness-raising campaigns run by the State and

---

[482] See the document entitled <u>Enquete post campagne sur 1 utilisation des MILDA 2016/2017. Rapport final</u>. Op. cit... p. 36.

[483] Every Thursday, between 6am and 10am, public spaces are cleaned. As a result, the steps, shops and shops don't open.

[484] See <u>Plan strategique national de lutte contre le paludisme au Cameroun 2007-2010</u>. Op. cit... p.56.

NGOs, were designed to encourage nationals to adopt and take on board malaria prevention measures. Clearly, these awareness campaigns have borne fruit, as many people in Central Cameroon are now using mosquito nets and insecticides, and many women are taking IPT.

The use of mil is the main health intervention implemented to reduce malaria transmission in Cameroon. Since 2003, MINSANTE has been distributing millet to households in accordance with the *"roll back malaria"* programme adopted by the government in 2002. In the Centre region, the use of ITNs/MILDAs by the population has improved considerably. Mosquito nets are being used more and more, as can be seen from the following table° 23:

Table no. 23

Use of mosquito nets by people in the

Centre-Cameroon region

after the 2016 LLIN campaign

Percentage of households owning at least one LLIN	Percentage of households with at least one mosquito net per sleeping space	Percentage of LLINs used for sleeping at night after the 2016 campaign	Percentage of children under five who slept half the night under an LLIN	Percentage of pregnant women aged 15-49 who slept under an LLIN last night
70.8% in Yaounde	53.9% in Yaounde	85% in Yaounde	71.5% in Yaounde	83.5% in Yaounde
78.7% at La Peripherie	41.6% at La Peripherie	70.4% at La Peripherie	63.2% at La Peripherie	76.1% at La Peripherie

Source : compiled by 1 author from data taken from Rapport final, MINSANTE, PNLP, INSdedecembre 2017,Op. cit... pp. 43-67.

It should be noted that the figures in this table are based on a survey carried out in 2016 and 2017 to evaluate the 2016 LLIN distribution campaign[485] . It is clear that the use of mosquito nets by people in the Centre-Cameroon region is well above average. For some sections of the population, the net impregnation rate is even well above the target set by the NMCP, which is 60%. In the city of Yaounde and the outskirts, the results are clearly satisfactory, even if the percentage of households with at least one mosquito net per sleeping space is still below average, especially in the outlying districts. This means that many people are still sleeping without a net in the Centre region. However, the results for children under five and pregnant women are satisfactory, with 71.5% of children under five and 83.5% of pregnant women in Yaounde sleeping under a mosquito net since the end of the 2016 campaign. In the peripheral zone, although the percentage is low, it is still appreciable, as here 63.2% of children under five and 76.1% of pregnant women have slept under a mosquito net since the end of the 2016 campaign. Tables° 24 and

---

[485] See the document entitled <u>Enquete post campagne sur 1 utilisation des MILDA 2016/2017. Rapport final</u>. Op. cit... pp. 43-67.

167

25 below provide a clearer picture of changes in the rate of net impregnation among these social strata.

Table 24

Percentage of children under five who slept under an LLIN last night

Periods	Before the mass distribution of LLINs (2014)	After the mass distribution of LLINs (2017)
Percentage of children under five who slept under a Mil	54,8%	63,0%

Source : compiled by 1 author from data taken from Rapport final, MINSANTE, PNLP, INSdedecembre 2017,Op. cit... pp. 43-67.

This table shows that the LLIN impregnation rate increased significantly for children under five between 2014 and 2017, rising from 54.8% to 63.0%. Empirically, this means that household behaviour in terms of receiving and using LLINs has changed significantly. Parents of children under the age of five are increasingly inclined to have their children sleep under a mosquito net. The situation is virtually the same with regard to the impregnation and use of LLINs by pregnant women.

Table 25

Percentage of pregnant women aged 15-49 who slept under an LLIN last night

Periods	Before the mass distribution of LLINs (2014)	After the 2017 mass distribution campaign for LLINs
Percentage of pregnant women aged 15-49 who slept last night under a Mil	52,3%	76,1%

Source : compiled by 1 author from data taken from Rapport final, MINSANTE, PNLP, INSdedecembre 2017,Op. cit... pp. 43-67.

It is clear from these two tables that the rate of impregnation with mosquito nets has increased among children under five and pregnant women since the launch of the 2016 mass LLIN campaign. Before the 2016 campaign, the percentage of children under the age of five who slept under an LLIN the last night was 54.8%, but after 2016, this rate rose to 63.0%. The percentage of pregnant women rose from 52.3% before the 2016 campaign to 76.1% after it.

Analysis shows that the various awareness campaigns that preceded the distribution of LLINs to households had a significant impact on the behaviour of the target groups. Indeed, the population has finally internalized that the use of mosquito nets is the principal means of reducing malaria transmission, especially among children under the age of five and pregnant women. The *"communication for behaviour change"* programme, the actions of political elites such as the Honourable Marie-Rose NGUINI EFFA and the major *"nightwatch"* campaign, have really borne fruit by having an impact on the behaviour of the target groups, who are increasingly inclined to get hold of an LLIN and use it, as can be seen from the above tables, despite the pockets of resistance that remain.

Still on the subject of vector control, there is a clear and growing trend towards the use of insecticides in the Centre region. Household insecticide spraying (HIS), which is an effective way of preventing the risk of malaria, is becoming more and more common, and is now even part of the *social routine of* people living in rural areas. A study carried out in the town of Ntui showed that out of fifty (50) households, forty-five (45) sprayed their homes with insecticide around twenty hours before going to sleep[486] . Shop assistants admit that insecticides are among the products most commonly sold to the local population[487] . The trend is not as strong in urban areas, since a study shows that out of fifty (50) households in the Biyem-Assi district, twenty-eight (28) spray their homes with insecticides before going to bed[488] . There is also widespread use of a non-spray insecticide called *"moon tiger"*, which is now being sold in the region like hotcakes. *In short,* the scrupulous respect shown by the local population for vector control measures is sufficient proof that the various behaviour change communication campaigns have had an impact on the target groups.

Furthermore, chemoprevention is now based on the use of IPT by pregnant women. In the social health field of Centre-Cameroon, there has been a marked increase in the number of pregnant women receiving IPT. In the city of Yaounde, seven (07) out of ten (10) women admit to having received at least three free doses of IPT in health facilities during pregnancy in the course of 2020. However, only two (02) women out of ten (10) acknowledge having received them in 2010[489] . From this we can deduce that pregnant women in Yaounde are increasingly inclined to undergo sulfadoxine-pyrimethamine-based IPT. This means that the various

---

[486] Study carried out by ourselves in January 2021 among a random sample of fifty households in the town of Ntui.

[487] After speaking to three shop managers in the town of Ntui, they all agreed that insecticides are very popular with the local population. They sell an average of four a day.

[488] Study carried out by ourselves in January 2021 among a random sample of fifty households in the Biyem-Assi district of Yaounde.

[489] Figures obtained from a survey of a sample of ten women in the city of Yaounde with at least three children in July 2021.

awareness campaigns carried out by health facilities and NGOs/associations during home visits have had a considerable impact on pregnant women.

In short, it can be said that the dynamics of the institutionalization of the model fight against the malaria problem has had real effects on the target groups which, through their behaviour, are supposed to be responsible for the spread of the disease. The effects on these groups can be measured in terms of the impact on their behaviour, which has improved considerably. In this case, the change in attitude is reflected in the intensification of hygiene practices and the adoption of prevention measures. It can therefore be asserted that the intensification of the fight against malaria from 2002 onwards has been effective and efficient at a social level in terms of its effects on the part of the population of Centre-Cameroon referred to as the target groups. These target groups now pay close attention to hygiene and malaria prevention measures. In addition to this change in people's attitudes, the process of institutionalising malaria control has had another major effect: it has improved the socio-economic conditions of households.

B- Improving the socio-economic conditions of households

The process of institutionalising the fight against malaria has not only had an impact on the individuals whose behaviour is at the root of the spread of the disease (target groups). It has also had a significant effect on the people directly affected by the malaria problem, known as *end* beneficiaries. End beneficiaries are the individuals (natural or legal) and their associations who are directly affected by the collective problem and who suffer the negative effects. As a result of the effective implementation of a policy, these actors can expect to see a (possible) improvement in their economic, social, professional, ecological or other conditions. The final beneficiaries are the actors who benefit, in a more or less direct way and depending on the objectives pursued by the policy in question, from the change in behaviour of the target groups[490] . *Ultimately,* these are the social players whose situation must imperatively improve as a result of the implementation of a social policy.

In the geographical area of Centre-Cameroon, certain people, by their behaviour, are considered to be responsible for the spread of malaria, while others suffer the immediate negative effects. Those who suffer the immediate negative effects are either people who actually fall ill with malaria, or the families and close friends of these people, who suffer the financial and moral consequences. The intensification of the modern fight against malaria is supposed to have an effect on the final beneficiaries, who should see their health and socio-economic situation improve, because its ultimate objective is to reduce the number of cases of malaria-related illness. Consequently, the effects of institutionalising the fight against malaria can no longer be assessed in terms of impacts, but in terms of social *outcomes.* To assess the effects (*outcomes)* of the tightening of malaria policies on the final beneficiaries in Centre-Cameroon, a series of questions need to be asked: have cases of malaria decreased in the population? Have pregnant women and children

---

[490] Read KNOEPFEL (P), LARRUE (C), VARONE (D) and HILL (M), <u>Public policy analysis.</u> Op. cit... p. 53.

under the age of five seen their situation improve? Has the financial burden of malaria on the family dropped significantly?

While the intensification of the fight against malaria in 2002 may have had a significant impact on the behaviour of the target groups, as shown above, it will also have a *de facto* effect on the final beneficiaries, since the situation of the latter depends in reality on the change in attitude of the target groups. As hygiene and prevention measures are now scrupulously respected by a large proportion of the population, malaria-related problems are likely to diminish naturally. Thus, since the implementation of the *"roll back malaria"* programme in 2002, and the various actions that have been taken since then, the number of people affected by malaria has risen sharply.

As a result, the number of cases of malaria reported to hospitals and the rate of malaria-related illness have fallen considerably **(1)**. **At the** same time, the financial costs associated with malaria have fallen sharply **(2)**.

1. A significant drop in the number of malaria cases reported in hospitals and in the malaria mortality rate

Since the implementation of the *"roll back malaria"* programme in 2002 in the social health field of Centre-Cameroon, there has been a clear decline in the number of malaria cases reported in hospitals in Yaounde and in rural areas. To demonstrate that malaria cases are falling in Yaounde health facilities, we will use the figures obtained from the district hospitals of Biyem-Assi, Cite-Verte and Nkolbisson. Tables° 26, 27 and 28 below show the gradual decline in malaria cases reported in Yaounde hospitals since 2002.

Table 26

Number of malaria cases recorded per year by the

Biyem-Assi district hospital

since 2002

YEARS	NUMBER OF CASES OF MALARIA RECORDED
2002	2,847 CASES
2003	2,756 CASES
2004	2,153 CASES
2005	2,056 CASES
2006	1,987 CASES
2007	1,809 CASES
2008	1,772 CASES
2009	1,768 CASES
2010	1,732 CASES
2011	949 CASES
2012	934 CASES
2013	929 CASES

2014	918 CASES
2015	787 CAS
2016	769 CAS
2017	546 CASES
2018	531 CASES
2019	527 CASES
2020	493 CASES

Source: compiled by the author using data from the general surveillance of
l'Hopital de district de Biyem-Assi

This table shows that there has been a gradual decline in the number of malaria cases reported at Biyem-Assi district hospital, from 2,847 cases in 2002 to 493 cases in 2020. The scenario is virtually the same for the Cite-Verte district hospital.

Table 27

Number of cases of malaria recorded per year by the
district hospital of
la cite-verte since 2002

YEARS	NUMBER OF CASES OF MALARIA RECORDED
2002	2,265 CASES
2003	2.103 CASES
2004	1,987 CASES
2005	1,950 CASES
2006	1,915 CASES
2007	1,882 CASES
2008	1,867 CASES
2009	1,811 CASES
2010	1,783 CASES
2011	887 CAS
2012	843 CASES
2013	835 CAS
2014	821 CASES
2015	645 CAS
2016	628 CASES
2017	443 CASES
2018	403 CASES
2019	387 CASES
2020	365 CASES

Source: compiled by the author using data from the general surveillance of
l'Hopital de district de la cite-verte

This table shows a gradual and progressive decline in the number of malaria cases reported at the Cite-Verte district hospital, with 2,265 cases reported in 2002 compared with just 365 cases reported in 2020. The situation is virtually the same at Nkolbisson district hospital.

Table no. 28

**Number of malaria cases recorded per year by the
Nkolbisson district hospital
since 2002**

YEARS	NUMBER OF CASES OF MALARIA RECORDED
2002	2,734 CASES
2003	2,407 CASES
2004	2.113 CASES
2005	2,033 CASES
2006	1,896 CASES
2007	1,875 CASES
2008	1,860 CASES
2009	1,810 CASES
2010	1,778 CASES
2011	794 CASES
2012	783 CASES
2013	771 CAS
2014	763 CAS
2015	573 CAS
2016	549 CASES
2017	394 CASES
2018	320 CASES
2019	308 CASES
2020	287 CASES

Source: Compiled by the author using data from the general surveillance of the Nkolbisson District Hospital.

Reading the three tables above, it is clear that the number of malaria cases reported in Yaounde hospitals is clearly declining[491] . In 2002, the Biyem-Assi district hospital recorded a total of 2,847 cases of malaria, compared with 493 cases in 2020. The Cité-Vite district hospital recorded a total of 2,265 cases in 2002, compared with 365 in 2020. The Nkolbisson district hospital recorded 2,734 cases in 2002, compared with 287 cases in 2020. From this point onwards, we can see a drastic drop in the number of cases of malaria reported in Yaounde's health facilities. This is largely due to the actions of social and institutional players in the process of tightening up the modern fight against malaria. These actions have led to

---

[491] It should be pointed out that the three hospitals selected here represent a sample from which we can estimate that the regression is complete in all the health facilities in this geographical area. It would have been very tedious, if not impossible, to go round all the health facilities in the city of Yaounde.

173

people complying with hygiene and prevention measures, thereby limiting the spread of the disease. As a result, malaria cases are tending to decline in health facilities.

A reading of these tables also reveals that 2011 and 2017 were the years in which a "sharp" decline in malaria cases in hospitals was recorded. In other words, it was during these years that Yaounde hospitals experienced a sharp drop in the number of malaria cases reported in health facilities. This shows that the campaigns described as "mass distribution of LLINs" had a significant influence on the risk of malaria in the city of Yaounde. To take the example of the Biyem-Assi district hospital, while 1,732 cases of malaria were reported here in 2010, only 949 cases were recorded the following year, i.e. in 2011! While 769 cases of malaria were reported in 2016, 546 cases were recorded in 2017. The rate of decline has suddenly become "rapid". The situation in district hospitals in rural areas is virtually the same, but with a slightly slower rate of decline.

As was done for the city of Yaounde, we will use a sample of three hospitals to show that reported cases of malaria have clearly declined in district health facilities in rural areas since the *roll back malaria* programme was adopted and implemented around 2000. Tables° 29, 30 and 31 show the decline in malaria cases in rural hospitals based on figures obtained from the district hospitals of Ntui, Akonolinga and Yoko.

Table 29
Number of malaria cases recorded per year by
Ntui district hospital
since 2002

YEARS	NUMBER OF CASES OF MALARIA RECORDED
2002	3421
2003	3298
2004	3114
2005	2995
2006	2916
2007	2884
2008	2736
2009	2711
2010	2634
2011	1205
2012	1088
2013	986
2014	967
2015	854
2016	797

2017	745
2018	703
2019	645
2020	577

Source: Compiled by the author using data from the general surveillance of the Ntui District Hospital.

From this table, it is clear that reported cases of malaria are clearly declining at the Ntui district hospital: in 2002, this hospital recorded 3,421 cases of malaria, compared with just 577 cases in 2020. The situation is virtually the same at Akonolinga district hospital.

Table 30

Number of malaria cases recorded per year by

Akonolinga district hospital

since 2002

YEARS	NUMBER OF CASES OF MALARIA RECORDED
2002	2674
2003	2611
2004	2589
2005	2575
2006	2502
2007	2478
2008	2348
2009	2298
2010	2109
2011	1278
2012	985
2013	974
2014	925
2015	801
2016	721
2017	703
2018	679
2019	511
2020	432

Source: Compiled by the author using data from the general surveillance of the Akonolinga District Hospital.

The table shows that the Akonolinga district hospital is experiencing a gradual decline in the number of malaria cases reported. In 2002, this hospital recorded 2,674 cases of malaria, compared with just 432 in 2020. The same trend can be seen at Yoko district hospital.

Table 31
Number of malaria cases recorded by
Yoko district hospital
since 2002

YEARS	NUMBER OF CASES OF MALARIA RECORDED
2002	1798
2003	1774
2004	1725
2005	1702
2006	1689
2007	1667
2008	1636
2009	1623
2010	1584
2011	907
2012	873
2013	859
2014	834
2015	801
2016	776
2017	720
2018	674
2019	594
2020	485

Source: Compiled by the author using data from general surveillance at Yoko District Hospital.

The above tables show that the number of malaria cases reported is also falling in rural hospitals, but at a "slow" rate. At Yoko district hospital, for example, 1,798 cases of malaria were recorded in 2002, compared with 485 cases in 2020. This means a drop of only 1,313 cases between 2002 and 2020, i.e. eighteen (18) years later! which is not a very flattering result compared to what was recorded in Yaounde district hospitals. This is largely due to the many tribulations that have significantly hampered the action of the social actors involved in the process of stepping up the fight against malaria in rural health districts.

In short, whether in the hospitals of Yaounde or in the outlying areas, the dynamic of institutionalising the fight against malaria has indeed had an effect, because the curve of declared cases of malaria is essentially regressive. It can be seen that the rate of regression is "rapid" in the Yaounde health facilities and "slower" in those on the outskirts. In all cases, however, the number of malaria cases recorded in hospitals tends to fall over the years. The actions of the collective and individual players responsible for stepping up the fight against malaria have ultimately had an impact on the social field. This can be seen from the malaria rate in the region between 2002 and 2020.

Since the adoption and implementation of the *"roll back malaria"* programme in 2002, there has been a gradual decline in the malaria rate in the Centre region. This can be seen in Tables 32, 33, 34 and 35 below.

Table 32

Number of deaths due to malaria per year in the Centre region since 2002

YEARS	NUMBER OF DEATHS RECORDED
2002	458
2003	397
2004	412
2005	320
2006	409
2007	343
2008	410
2009	415
2010	404
2011	155
2012	235
2013	249
2014	229
2015	304
2016	203
2017	140
2018	190
2019	210
2020	296

Source: compiled by the author using data from the health information and planning department of the Delegation Regionale de la Sante Publique du Centre.

The table shows that the rate of malaria is falling in the Centre region. In other words, the number of deaths due to malaria is falling: in 2002, the region recorded 458 deaths due to malaria, compared with 296 in 2020. It should be noted, however, that this fall in the malaria death rate was greater in 2011 and 2017, the years corresponding to the mass distribution campaigns for LLINs. Apart from these two years, the fall in the malaria death rate is essentially relative. As a result, it is becoming necessary, and indeed imperative, for the public authorities to step up mass distribution campaigns for LLINs.

Table no. 33

Number of deaths due to malaria per 100,000 inhabitants between 2014

and 2018 in the Centre region

YEARS	2014	2015	2016	2017	2018
Number of deaths/100,000 inhabitants	6	4	8	6	5

Source: <u>Rapport de suivi des 100 indicateurs cles de sante au Cameroun en 2019, focus sur les ODD,</u> MINSANTE-OMS, ONSP, 2019, p. 47.

According to the table above, for a population of 100,000 in the Centre region, the number of deaths due to malaria fell slightly between 2014 and 2018. For every 100,000 inhabitants, there were six (06) deaths in 2014 and five (05) deaths in 2018. The difference between the two years is not as significant.

Table 34

Number of deaths due to malaria in children under five since 2002 in the Centre region

YEARS	NUMBER OF DEATHS RECORDED
2002	283
2003	244
2004	255
2005	249
2006	253
2007	202
2008	117
2009	263
2010	235
2011	96
2012	145
2013	154
2014	132
2015	188
2016	102
2017	81
2018	96
2019	113
2020	197

Source: compiled by the author using data from the health information and planning department of the Delegation Regionale de la Sante Publique du Centre.

The table above shows that the number of deaths due to malaria in children under the age of five fell "slightly" between 2002 and 2020, from 283 to 197. In other words, the drop in the rate of malaria in children under five is not as great as the "*Roll Back Malaria*" programme would have liked. We can also see that 2011,

178

2017 and 2018 were the years in which the region recorded the lowest malaria-related death rates among children under five, with 96, 81 and 96 deaths respectively. These were the years that followed directly on from those devoted to "mass" LLIN distribution campaigns across the region. Hence the need to multiply this type of campaign in order to contain, if not curb, the threat of malaria in the social health field of Central Cameroon.

Table 35

Number of deaths due to malaria among pregnant women since 2002 in the Centre region

YEARS	NUMBER OF DEATHS RECORDED
2002	21
2003	19
2004	20
2005	11
2006	20
2007	23
2008	21
2009	20
2010	23
2011	08
2012	12
2013	11
2014	10
2015	15
2016	09
2017	06
2018	12
2019	15
2020	16

Source: compiled by 1 using data from the information service
of the Delegation Regionale de la Sante Publique du Centre (Regional Delegation for Public Health in the Centre)

The above tables show that the malaria mortality rate is declining in the Centre's social health field. Between 2002 and 2020, deaths due to malaria fell considerably among children aged between zero and five, pregnant women and the rest of the population suffering from malaria. In 2002, the Centre region recorded 458 deaths from malaria, including 283 children under the age of five, 21 pregnant women and 154 deaths outside these two social strata. In 2020, the region recorded 296 deaths from malaria, i.e. 197 children under the age of five, 16 pregnant women and 83 registered deaths outside the two social strata. From 2002 to 2020, the decline in malaria deaths is clearly visible.

The tables also show that 2011 and 2017 were the years in which the region recorded the lowest rates of mortality, with 155 deaths in 2011 and 140 in 2017. Children under the age of five suffered 96 deaths in 2011 and 81 in 2017. Pregnant women suffered 08 deaths in 2011 and 06 in 2017. It's easy to see that the years with the lowest malaria mortality rates were the years following the mass distribution of mosquito nets to households. These campaigns, because they were accompanied by extensive awareness-raising campaigns, considerably reduced the impact of malaria on the population. Hence the need for the public authorities to multiply this type of campaign in order to definitively reduce the impact of malaria on the population. Unfortunately, unlike the other regions of Cameroon, the Centre did not see any mass distribution of LLINs in 2019, and as a result, in 2020, the region recorded 296 deaths from malaria, 156 more than in 2017. This situation has seriously altered the downward trend in the malaria mortality rate that began in 2017. But the least we can say is that the implementation of the *"roll back malaria"* and *"K.O malaria"* programmes has had a significant effect on the final beneficiaries, who have seen their situation improve with the general drop in the malaria death rate. The *results of implementation* have also been perceptible on the social front, even though the problem of malaria still persists, given the figures recorded in 2020. Finally, the malaria death rate is very high among children under five, compared with other sections of the population. In fact, 62% of deaths are among children under five, 5% among pregnant women and 33% among the rest of the population. This can be seen in graph° 1 below.

Graph 1
Graphical representation of the malaria mortality rate in the
Centre region
Malaria according to social class

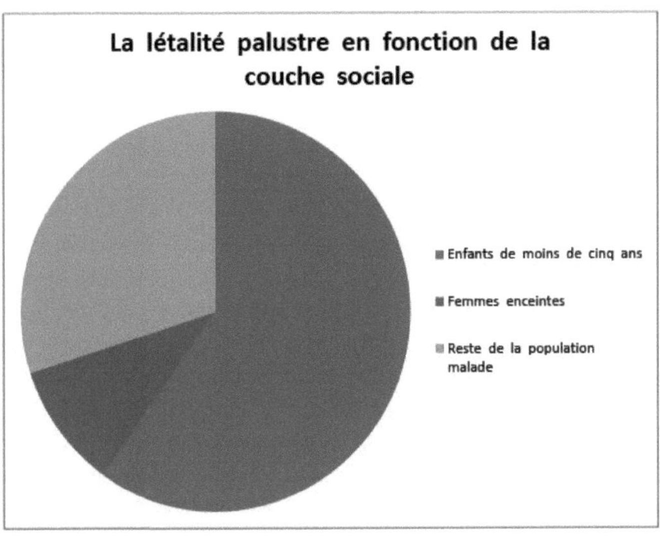

- Children under five years of age

- Pregnant women

- Rest of the sick population

An additional effort must be made by social and institutional players to target children under five and pregnant women, as it is these groups that have the highest rate of malaria in the region. It should also be pointed out that children under five and pregnant women represent a total of 22% of the population of the Centre, with the highest risks of morbidity and mortality. Consequently, if we want to significantly reduce the rate of malaria in this social sector, we need to maximise social actions in favour of these social groups.

In addition to the significant drop in the number of malaria cases reported to hospitals and the malaria mortality rate, the drive to institutionalise modern malaria control is having another major impact on households: lower healthcare spending.

2. The fall in household healthcare expenditure

As the final beneficiaries have seen their situation improve as a result of the intensification of the modern fight against malaria, those who are directly close to them will benefit indirectly from this situation: these are the *positively affected third parties*. As malaria declines considerably, as shown above, the financial burdens on households associated with malaria will also fall considerably. In fact, when an individual in a family suffers from malaria, it is the whole family that is actually affected because a financial effort will have to be made to "save" the patient. In the long term, this can lead to serious poverty. However, if the number of sick people in households decreases, the financial situation of families will be restored. According to Joseph NTANGSI's study, household healthcare expenditure breaks down as follows:

- Pharmacies and other drugstores: 53.0%;
- Private not-for-profit training: 23.1%;
- Public health training: 14.6% ;
- Traditional medicine: 7.0% ;
- For-profit clinics: 2.3%[492] .

In 2010, each Cameroonian household spent an average of 83,400 CFA francs (annually or monthly?) on healthcare, or 13,900 CFA francs per person for an average family of six[493] . Seventy percent (70%) of this expenditure was directly linked to the problem of malaria. Today, the trend is considerably downwards with the adoption of hygiene and malaria prevention measures by the population. In the Centre region, a survey carried out in January 2020 revealed that each household spends an average of 30,000 FCFA a year on healthcare, or 5,000 per person for an average family of six[494] . This represents a clear improvement on the situation in 2010. In other words, financial expenditure on healthcare is tending to fall over the years. Since malaria accounts for more than half of all cases of illness recorded in hospitals (70% according to public health professionals)[495] , it can be assumed that the majority of healthcare expenditure is due to malaria. It is also easy to assume that the fall in healthcare expenditure is directly linked to the fall in malaria cases recorded in health facilities (see *above).

In short, the drive to institutionalise the fight against malaria has had a significant effect on the end beneficiaries, who have seen their health situation improve with a drastic fall in the number of malaria-related deaths. In contrast to the situation at the beginning of the 2000s, malaria mortality has now fallen considerably, according to the figures in the tables above. A fall in the malaria rate also means a fall in the number of malaria cases. This could not be any other way, given the increase in compliance with hygiene and malaria prevention measures. The change in the behaviour of the target groups has led to an improvement in the situation of the final beneficiaries and third-party beneficiaries such as households and families, who have seen their healthcare expenditure drop considerably. Done, the tightening of malaria policies in 2002 produced effects (impacts+outcomes) firstly on the target groups (impacts) and then on the final beneficiaries (outcomes) in the Centre-Cameroon region. In this respect, it can be said to have been effective, efficient and relevant, even if much remains to be done, as the disease persists.

Paragraph 2: The persistence of the malaria problem

It would be illusory to think that the malaria problem had completely disappeared

---

[492] See <u>Plan Strategique National de Lutte contre le Paludisme du Cameroun 2007-2010</u>. Op. cit... p. 28.
[493] Ibid. p. 28.
[494] Survey carried out in January 2020 among a sample of fifteen households: five in a poor neighbourhood (Elig-Dzoa), five in an intermediate neighbourhood (Ngousso) and five in a rich neighbourhood (Bastos). The survey showed that health expenditure was higher in well-off households and much lower in poor households.
[495] "Out of ten in-patients here, seven suffer from malaria", said Nathalie A., a nurse at a hospital in the Tsinga district, as reported by Rabiatou NANA in the *Cameroon Tribune* on 24 September 2020.

following the intensification of the fight against malaria in 2002, because despite the many actions undertaken by both collective and individual actors, both institutional and social, the disease persists in the social health field of Central Cameroon. In fact, the many ups and downs in the process of institutionalising the fight against malaria, as described in the previous chapter, have had a serious impact on the activities of the implementers in the geographical area of Centre-Cameroon. In reality, a public policy is never implemented in a 'vacuum'; it is always introduced into a context characterised by problems, structures and constraints that predate it, which provokes reactions in return. These reactions in turn influence the activities of the implementers[496] . This is why the problem of malaria still persists in the Centre-Cameroon region, despite the considerable drop in reported cases of malaria in health facilities, as shown above. To carry out our analysis, we will rely on a quantitative analysis based on statistics obtained in 2020 in hospitals in the Centre-Cameroon geographical area (A). It should also be noted that the persistence of the malaria problem is also reflected empirically in patients' constant recourse to parallel pharmacies and traditional practitioners (B).

A- Persistence in hospitals

Although it is clearly declining in 1 geographical area of Centre-Cameroon, the malaria problem still persists in this region. The real empirical reason for this is that, in 2020, the hospitals in this social health field recorded cases of malaria, broken down according to the following tables° 36, 37, 38 and 39:

Table 36

Cases of malaria recorded at Biyem-Assi district hospital in 2020 broken down by month

MONTHS	NUMBER OF PATIENTS REGISTERED
January	42
February	41
March	86
April	43
May	42
June	45
July	44
August	43
September	41
October	23
November	22
December	21
TOTAL	**493 CASES**

Source: Compiled by the author using data from the general surveillance of the Biyem-Assi district

---

[496] See KUBLER (D) and MAILLARD (J-D), <u>Analyser les politiques publiques</u>. Loc. cit... p. 73.

hospital.

The table shows that there were still many cases of malaria reported in the Biyem-Assi health district in 2020, with 493 cases recorded.

Table 37

Cases of malaria recorded at the

Cite-Verte district hospital

in 2020, broken down by month

MONTHS	NUMBER OF PATIENTS REGISTERED
January	32
February	34
March	67
April	35
May	28
June	25
July	31
August	32
September	33
October	17
November	15
December	16
TOTAL	**365**

Source: Compiled by the author using data from the general surveillance of the Cite-Verte district hospital.

The table shows that there were still many cases of malaria reported in the Cite-Verte health district in 2020, with 365 cases recorded.

Table 38

Cases of malaria recorded at Ntui district hospital in 2020

broken down by month

MONTHS	NUMBER OF PATIENTS REGISTERED
January	56
February	56
March	80
April	54
May	49
June	53
July	49
August	54
September	52

October	27
November	22
December	26
TOTAL	**577 CASES**

Source: Compiled by the author using data from the general surveillance of the Ntui district hospital.

The table shows that there were still many cases of malaria reported in the Ntui health district in 2020, with 577 cases recorded.

Table 39

Cases of malaria recorded at Yoko district hospital in 2020 broken down by month

MONTHS	NUMBER OF PATIENTS REGISTERED
January	52
February	54
March	94
April	50
May	43
June	47
July	38
August	31
September	29
October	19
November	13
December	15
TOTAL	**485**

Source: Compiled by the author using data from general surveillance at Yoko district hospital.

A reading of the four tables above, which present the situation of malaria cases recorded in 2020 both in hospitals in Yaounde and in those on the outskirts, shows that malaria cases have again been legion in the geographical area of Central Cameroon. The month in which the most cases of malaria are recorded in the region's health facilities is March. In March 2020, the Biyem-Assi and Cite-Verte district hospitals in Yaounde recorded 86 and 67 cases of malaria respectively. In the same month of the year, the district hospitals of Ntui and Yoko, on the outskirts of Yaounde, recorded 80 and 94 cases of malaria respectively. On the other hand, November was the month in which the region's hospitals recorded the fewest cases of malaria. In fact, in November 2020, the district hospitals of Biyem-Assi and Cite-Verte recorded 22 and 15 cases of malaria respectively. The rural hospitals of Ntui and Yoko recorded 22 and 13 cases of malaria respectively in November 2020.

The analysis that can be made from reading these tables is that the populations of the Central region are more exposed to the risk of malaria during the months of January, February and March; this is because these months correspond to the dry season with high night-time temperatures and are characterised by the absence of rain. This encourages mosquitoes to hatch and spread. On the other hand, the months of October, November and December are the months when people are least exposed to the risk of malaria. This is because these months are characterised by heavy rainfall and, above all, violent cold at night, which considerably limits the development and deployment of mosquitoes. This is why, when we look at the four tables above, the figures for the first three months are the highest and those for the third three months the lowest. Therefore, public and private players must do more to run public awareness campaigns at the start of each year to reduce the high incidence of malaria during the first three months.

In the final analysis, the four tables show that malaria remains a risk in the region. Although the number of cases reported in hospitals has fallen, the risk has not disappeared. The fact that the Ntui district hospital still recorded 80 cases of malaria and the Biyem-Assi hospital 86 cases in March 2020 means that the threat persists, and much remains to be done to definitively curb the disease. It should also be noted that people in the Centre-Cameroon region continue to use the informal healthcare system and parallel pharmacies to treat malaria. This is further evidence that malaria still persists in the region.

B- The persistence of patients' recourse to the informal healthcare system and to parallel pharmacies

The main social response to the breakdown of the conventional healthcare system has been the emergence of a powerful informal healthcare system[497] . This system has three main components, including traditional healthcare and prophetic healthcare[498] . In the geographical area of Centre-Cameroon, herbalists are increasingly sought after by patients. In other words, apart from the hospital, many households still seek the expertise of traditional doctors for treatment. In Yaounde, the *"modern traditional clinic"* run by Dr DEWAH and BROS and the "BK traditional clinic" receive a large number of severe malaria patients every week. A herbalist at Dr DEWAH's clinic justified this when he said: "*We see a lot of people with severe malaria here. They come here when òa has passed them by at home. We see an average of three patients a week"[499]* . Doctor BK (Brillant Kamdem) admits that he receives a large number of severe malaria patients in his traditional clinic in the Etoudi neighbourhood, at a place called ancien sixieme in Yaounde. On the Mokolo market, a famous naturopath claims to receive malaria patients and treat them with artemisa. Dr Moherb's NOKADJI has several branches in Yaounde, the best known of which is located in the Mokolo district. "*I always see people with*

---

[497] Read MEVA'A ABOMO (D), "Le fardeau de la lutte contre le paludisme urbain au Cameroun : etat des lieux, contraintes et perspectives", Op. cit... p. 35.
[498] Ibid. p. 35.
[499] Interview with a doctor at Doctor DEWAH and BROS's clinic in Yaounde on 16 June 2021.

malaria here, and I treat them with artemisia" .[500]

Products made from the Artemisia plant and packaged in sachets are currently selling like hotcakes in Yaounde. The product in question is "ARPACURE", concocted by a group of Cameroonian herbalists and used mainly to treat malaria. It costs between FCFA 1,000 and FCFA 2,000. It is consumed in the form of a hot herbal tea and has unimaginable virtues, according to the people who use it. Some households say they buy four sachets of "ARPACURE" a week to treat themselves at home and even to prevent the risk of malaria.

In rural districts, traditional medicine has been very successful, due to the "crisis in conventional healthcare provision"[501] . In most cases, the herbalists take care of the sick in their own homes. They are even the real driving forces in the fight against malaria in the village health areas. In the Yoko health district, for example, a very popular herbalist by the name of Joseph Sidi treats malaria patients at his home in the Yoko district, using medicinal plants. The plant known as "artemisia" is the resource Joseph SIDI uses to "tame" malaria in the Yoko sub-system. The treatment procedure is described by a patient in the following terms:

"When you go to see Sidi when you're suffering from malaria, he asks you to sit in his small living room while he takes a look around a room in his flat. Two minutes later, he comes out with a bottle of water and some plants. He boils the water in a pot for at least fifteen minutes with the plants inside, then filters it through a very fine sieve and pours the contents into a bottle to drink morning and night"[502] .

The success of the traditional pharmacopeia practised by SIDI in the social health field of Yoko comes from the fact that it is less costly and quicker in treatment than what is done in hospitals. Some patients go for one reason or the other, or both. But for the majority of people living in Yoko, it is the cost of malaria treatment by Joseph SIDI that motivates them to go to him rather than to the hospital. In the Ntui health district, another very popular herbalist occupies social space in the fight against epidemics. Called *"Docteur Federal"* or *"Le Naturopathe",* he operates differently from SIDI in that he takes care of the sick person at home. The patient calls *"Docteur Federal"* to explain his or her problem, and the doctor immediately goes to the patient's home with the necessary remedy, which is always in liquid form and contained in a bottle of water. The effectiveness of this traditional product and its low price encourage people to call on this herbalist rather than a nurse. The words of the traditional practitioner known as *"Docteur Federal"* show just how popular traditional pharmacopeia is with the people of Ntui, when he says that "on *average, I receive fifteen calls a day from sick people"[503]* .

Prophetic medicine also plays an important role in the fight against malaria in rural areas of the Centre region. The pastors of the so-called "revivalist" churches are

---

[500] Interview with Dr Moherb's MOKADJI in Mokolo on 15 June 2021.

[501] People are increasingly losing confidence in doctors because of the corruption, swindling and false diagnoses they practise. What's more, the dilapidated state of some village health facilities is not encouraging people to go there.

[502] Interview with Simeon, a native of Yoko, on 9 March 2021.

[503] Interview with 1 herbalist known as "Docta" or "Doctor Federal" on 31 March 2021.

increasingly called upon by the local population to solve their health problems. The free prophetic medicine practised by these pastors means that they are much more in demand in the event of illness than nurses and herbalists. When it comes to malaria, the method of treatment is no different from that for other diseases, and is based on prayer. The "laying on of hands"[504] of the pastor on the patient's head has therapeutic virtues, if we are to believe the words of a citizen of the town of Ntui:

"I had a violent headache once, accompanied by a fever. I knew it was malaria. Fortunately, that day was a Friday and it was prayer night at church. That evening, I explained my problem to the pastor, he laid his hands on my head and started praying. And I can assure you that the next morning, I had neither a headache nor a fever. I was completely healthy"[505] .

From these words, it is clear that prayer has therapeutic virtues, that it effectively treats the patient. The symptoms of malaria diminish after the patient has been exposed to a session of laying on of hands. This sociomedical act is seen here as the most effective treatment for malaria, which is seen as a metasocial attack[506] . With poverty in full swing in the villages of the Centre region, prophetic medicine, because it is free, has become the population's bulwark for health care. From this angle of analysis, diseases in general and malaria in particular seem to have become pathologies stemming from "malefic" or "diabolic" forces. In fact, this is what pastors tell their followers, that "all illnesses are caused by the devil" and that "deliverance is the right remedy". All of which leads a large proportion of the population to turn to them in the event of malaria. *Ultimately,* in the districts on the outskirts of the Centre, the informal health care system has practically supplanted conventional medicine. However, it should be pointed out that traditional healthcare is recognised and encouraged by the public authorities, unlike prophetic healthcare, which remains "clandestine". Despite the modernisation of the fight against malaria, so-called traditional and archaic practices still persist in the Centre region: this is the phenomenon of *Path Dependence.*

When talking about parallel pharmacies, it should be pointed out that these refer to the clandestine and informal structures used to treat sick people, as mentioned in the previous chapters. In fact, many people continue to use this type of pharmacy in cases of simple or serious malaria, despite the fact that the public authorities have repeatedly banned the illegal sale of medicines. In Yaounde, as elsewhere in the health sector of Central Cameroon, people who go to street drug sellers to buy anti-malarial drugs cite a lack of financial resources. The medicines sold on the street are cheaper and sometimes effective in treating the disease, people admit.

As a result, street medicines continue to be sold in the region, encouraged to some extent by local people who, for their own reasons, prefer to buy them rather than go to a conventional pharmacy. To this day, the "Bamenda boys" continue to receive people who ask them for anti-malarial drugs to ensure their recovery. In Yaounde,

---

[504] The pastor places his hands on the head of the kneeling patient and begins a prayer that lasts at least five minutes.
[505] Interview with a citizen of the town of Ntui, a follower of the church known as "the true Church of God in Cameroon", who did not want to reveal his identity.
[506] MOULIOM MOUNGBAKOU (I.B), Ibid. p. 151.

they are to be found at the central market, the Mfoundi market and Avenue Kennedy. In rural areas, they can be found in markets, sheds and ponds. Medicines are sold in the open air, sometimes on the ground. These street vendors make anti-malarial drugs available to patients at very low prices. The fact that people are resorting to the informal health care system and to parallel pharmacies is proof enough that the malaria problem still persists in the Centre region, despite the strong deployment of social and institutional players tasked with stepping up the fight against this disease. This state of affairs upsets the equilibrium sought in the process of institutionalising the fight against malaria: it is the phenomenon of "*punctuated* equilibrium".

The considerable drop in malaria cases recorded in health facilities does not mean that the malaria problem has disappeared. Admittedly, since the adoption and implementation of the *"roll back malaria"* programme in 2002, there has been a drastic reduction in malaria cases in hospitals, which is highly appreciable. However, in 2020, eighteen (18) years later, hospitals are still recording cases of malaria (see tables 38, 39, 40 and 41 above). It should also be noted that in 2020, people in the Centre region are still turning to herbalists, prophets and street vendors for treatment. Therefore, despite the fact that the dynamics of institutionalising the fight against malaria have considerably reduced the risk of this disease in the geographical area of Central Cameroon, it should be noted that the threat still persists today.

Although the problem of malaria is still very much in evidence in the geographical area of Centre-Cameroon, despite the dynamics of modernisation, it has to be said that this so-called "moderate" form of malaria control can be improved.

SECTION II

The perfectibility of modern malaria control

Despite the many trials and tribulations that have beset the drive to institutionalise modern malaria control in the Centre-Cameroon region, making its effects ambivalent, it should nevertheless be noted that this so-called "modern" form of malaria control can be improved. To achieve this, a number of actions need to be taken by the public authorities. They must step up sanitation campaigns, the distribution of LLINs and the fight against street medicines **(paragraph 1). In addition,** the public authorities must step up the fight against isolation and "medical deserts" and set up an operation to control the cost of malaria in health facilities **(paragraph 2).**

Paragraph 1: Stepping up sanitation campaigns, distribution of LLINs and the fight against the illegal trade in medicines

In order to improve the modern fight against the malaria problem in the geographical area of Centre-Cameroon, the public authorities must implement a certain number of actions, in particular: intensify the organisation of sanitation campaigns in the region's towns (A), organise recurrent campaigns for the distribution of LLINs and intensify the fight against the illegal trade in medicines

(B).

A- Intensify the organisation of sanitation campaigns in the region's towns and cities

As demonstrated in the previous chapter, the geographical area of Central Cameroon still has several "ecologically favourable zones"[507] and is faced with indescribable insalubrity. This state of affairs favours the development of the female anopheles responsible for malaria through the plasmodium it inoculates into the bloodstream. The public authorities must therefore take action to limit as far as possible these "ecologically favourable areas", which are present in almost all districts of the city.

Yaounde. In the Biyem-Assi district, the large marshy area known as "nage-glisse" needs to be cleaned up if we are to prevent mosquitoes from continuing to thrive there and whistling even in the middle of the day. To make this possible, sophisticated machinery is needed to clear and level the area. In the Elig-Edzoa district, the area known as the "bas-fond marche" needs special treatment from the Yaounde city authorities. This is a commercial area with a high concentration of people who are constantly in danger because of the unkempt buildings there. In the Mballa II district, some households still live on the edge of the yards, *as in* the Etoa-Meki district, to name but a few. All these "ecologically favourable areas" are prime breeding grounds for mosquitoes and need to be cleaned up by the public authorities if the fight against malaria is to be more effective in social terms.

The city of Yaounde also needs to be cleaned up by combating the insalubrious conditions that are rampant in the area. Household waste and manure still litter the alleyways under the gaze of the authorities, who blame the company in charge of waste management in the city, HYSACAM. The company's employees are constantly complaining about back pay. As a result, *Yaoundeans* are living with household waste and the risk of malaria is becoming high. The city's authorities must therefore assume their responsibilities, firstly by raising people's awareness of the need to dispose of household rubbish in the rubbish bins found at virtually every crossroads[508] . Secondly, by calling on HYSACAM to do its job and only its job. Finally, by paying the company's staff, who more often than not justify their inactivity with back pay. What's more, other waste management companies need to be set up in Yaounde to compete with HYSACAM, which sometimes finds itself overwhelmed by the workload, because the population of Yaounde has been growing exponentially for almost four years now, due to internal migration caused by the socio-political crisis in the North-West and South-West regions. If we are to improve the fight against malaria in the Centre region, we also need to organise additional campaigns to distribute LLINs and combat the illegal trade in medicines.

---

[507] These are areas where mosquitoes can spread and where the risk of malaria is consequently too high.

[508] It is worth noting, however, that despite the presence of rubbish bins, some people take a malicious pleasure in dumping rubbish not in the bin but outside it.

B- Organise additional campaigns to distribute LLINs and step up the fight against the illegal trade in medicines

In order to improve the fight against malaria in the Centre-Cameroon health sector, the public authorities need to step up campaigns for the "mass" distribution of LLINs in this area. As demonstrated above, it was following the State's mass distribution campaigns of LLINs to the population (particularly in 2011 and 2016) that the region saw a drastic drop in malaria-related morbidity and mortality. The process must therefore be stepped up if the region's authorities are to contain, if not curb, the threat of malaria. Unfortunately, the government's semi-annual mass distribution of LLINs to the population of the Centre region dates back to 2016. There were plans for another LLIN campaign in 2019, but to date this has never taken place, despite the fact that the populations of the nine other regions of Cameroon received free mosquito nets from the State on that date. The corollary of this de facto situation is that the malaria-related mortality rate rose considerably between 2019 and 2020. However, the death rate curve was considerably down in 2017. This means that campaigns to distribute LLINs to households are making a significant contribution to reducing the risk of malaria. It is therefore imperative for the State to organise a new campaign for the "mass" distribution of LLINs to the population in order to "fill the gap" left in 2019. The campaign should be stepped up in the city of Yaounde, where there are many ecologically favourable areas. But unfortunately, as we reach the beginning of 2021, there is still no sign of the slightest action, and the situation persists!

In addition, rationalising the fight against malaria also involves combating the illegal trade in medicines that persists in the city of Yaounde, where the institutions are based. Many people have recourse to medicines to ensure their recovery from simple or severe malaria. But not everyone can afford the quality of this medicine, as there is still a 'black market' in medicines in the region (this has been amply demonstrated and developed in the previous chapter). Peddlers of medicines are still rife on the streets, despite the fight undertaken by the Minister of Public Health, Mr MANAOUDA Malachie, who issued a statement on 10 July 2019 formally banning the sale of medicines from the informal market. MINSANTE agents and the Minister himself carried out raids to track down street vendors selling medicines of dubious quality. This was very much appreciated and welcomed. Unfortunately, this action was limited to Yaounde, the capital and headquarters of the institutions, to the detriment of rural areas. What's more, it didn't last because street drug sellers continue to flourish in the region y including in the city of Yaounde, putting many people at risk because fake anti-malarials are sold to them at very low prices!

Generally speaking, the institutions responsible for product quality control must also ensure the quality of the antimalarial drugs sold on the Cameroonian market, because, according to some, counterfeit medicines can be found even in so-called "conventional" pharmacies.

Paragraph 2: step up the fight against isolation and "medical deserts" and set up an operation to control the cost of malaria in health centres

In order to improve the way in which the malaria problem is tackled in the social health field in Centre-Cameroon, the public authorities must also implement a number of actions, in particular: combat isolation and build modem health centres **(A)**. They must also ensure strict compliance with the cost of malaria in health centres (B).

A- Stepping up the fight against isolation and "medical deserts

The twenty-four health districts in the rural areas of the Centre-Cameroon region are still home to a large number of highly sensitive areas.

enclaves. As we showed in the previous chapter, these areas are a serious obstacle to the fight against malaria. Awareness-raising campaigns, the distribution of LLINs and the treatment of patients at home are practically inoperative in these areas because they are so isolated. It is difficult to enter these rural areas either on foot or by car. The health districts of Ntui and Yoko in the department of Mbam-et-Kim are in reality forest areas in which communication routes are practically non-existent. This social reality is unsuitable for the harmonious implementation of programmes such as "home-based malaria treatment" (PECADOM), because penetrating the area is a major risk, given the state of the roads. How can a safe distribution campaign for home mosquito nets be carried out in areas   where

road traffic is practically non-existent? It is therefore up to   the   public authorities to open up these areas to ensure the proper implementation of modem malaria control strategies.

In the same vein, it has been observed that, unlike the city of Yaounde, the rural areas of the Centre region have a high incidence of "medical deserts". People in these areas are still sometimes obliged to travel twenty (20) or even eighty (80) kilometres before finding a health centre to receive treatment in the event of illness. This is the case in the Akonolinga health district, where the village of Akak has no quality hospital, and as a result, the people who live there are forced to travel forty kilometres to hope to find a hospital worthy of the name. This is also the case in the Ntui health district, where the villages of Ndimi, Ngoro and Ngambe-Tikar have no quality hospitals, meaning that people are forced to travel some twenty kilometres to reach Ntui in order to receive treatment in a hospital that is virtually state-of-the-art. Examples of this are legion. It is therefore up to the public authorities to provide rural areas with modem health centres so that sick people do not have to make long journeys each time to seek conventional health care, with all the risks that this entails. If the fight against malaria in the Centre region is to be perfected, it is imperative that an operation be set up to control the cost of malaria in health facilities.

B- Setting up an operation to control the cost of malaria in health centres

To make the fight against malaria more effective in the Centre-Cameroon region, the public authorities need to set up a "health watch" to ensure that the cost of treating malaria, as stipulated in the current regulations, is strictly adhered to. In several social areas in the region, the RDT for children under the age of five and the treatment of simple and severe malaria for children in this age group are still charged for. However, between 2010 and 2014, the President of the Republic decided to make them free. A laudable initiative, but one that unfortunately continues to be sidelined by health centre managers. At the Ntui district hospital (reputed to be the largest public health facility in the Mbam-et-Kim department), the RDT and the treatment of simple or severe malaria in children under the age of five are not free of charge, and no one is unaware of this. This is the case in all the health facilities in the Mbam-et-Kim department. This state of affairs hampers the institutionalisation of the fight against malaria in this area, because many people, already destitute, are still obliged to dig deep into their pockets to pay for treatment for their children under the age of five. As a result, some households are forced to resort to traditional and prophetic medicine. Worse still, they are forced to use parallel pharmacies, with all the damage that entails. The public authorities therefore urgently need to put in place a "health watch" so that people can fully benefit from the free malaria treatment that the Head of State wants. Sanctions must be applied against doctors who continue to charge households for malaria treatment for children under five. This will undoubtedly limit the spread of parallel pharmacies, which people currently use to ensure their health.

In conclusion, the effects of the process of institutionalising malaria control in the Centre-Cameroon region are clearly visible in the social arena. First of all, the people whose behaviour was required to change (the target groups) did in fact change their behaviour, which was at the root of the spread of the malaria problem, by adopting new attitudes aimed at promoting compliance with hygiene measures such as clearing drains, treating and cleaning latrines, and using improved water sources. These groups also scrupulously respect malaria prevention measures such as vector control and chemoprevention. In this way, LLINs/ITNs are increasingly used by households in all social categories. Pregnant women are increasingly inclined to use at least one IPT during their pregnancy.

The change in the behaviour of the target groups is leading to an improvement in the conditions of the final beneficiaries, who are increasingly seeing themselves as being free from malaria. In fact, since the *"roll back malaria" programme was adopted and implemented,* there has been a considerable drop in the malaria mortality rate in the social field of Central Cameroon, as the malaria deaths recorded in 2020 contrast sharply with those recorded in 2002. In other words, fewer and fewer people are dying of malaria in the Centre region. What's more, the number of malaria cases reported to hospitals has tended to fall over the years, as the number of patients is essentially decreasing. If we look at the curve of malaria cases reported in hospitals in Yaounde and the surrounding area between 2002 and 2020, we come to the conclusion that malaria is falling in the geographical area of Central Cameroon. As the malaria problem falls, so should the financial expenditure of households linked to this disease. This reduces the risk of poverty. So the *results of implementing* public malaria policies are clearly visible in the social arena. But this visibility of impacts and *outcomes* should not obscure the fact that the results obtained from the implementation of malaria policies are mixed or ambivalent.

In spite of the considerable drop in the number of malaria cases recorded in hospitals in Centre-Cameroon, it has to be said that the problem of malaria still persists in this area of health care. This persistence is reflected empirically by the fact that in the year 2020, health facilities in Centre-Cameroon still recorded high numbers of malaria cases. The persistence of the malaria problem is also reflected in the fact that people in this geographical area resort to informal healthcare services and parallel pharmacies. We therefore believe that the public authorities need to pay more attention to this public health problem, by stepping up both material and symbolic actions aimed at the population.

Although the problem of malaria is still very much in evidence in the geographical area of Centre-Cameroon, despite the dynamics of modernisation, it has to be said that there is still room for improvement in this so-called "modern" form of malaria control. To achieve this, the public authorities need to implement a number of measures both in the city of Yaounde and in rural areas. In the city of Yaounde, the public authorities must:

- cleaning up the city by combating environmentally unfavourable areas and insalubrious conditions;
- organise additional and regular campaigns to distribute LLINs;
- combat the sale of street drugs and counterfeit medicines by "tracking down" their promoters.

In rural areas, modern malaria control can be improved if the public authorities :

- combat isolation;
- build modem hospitals;
- to combat illiteracy;
- ensure strict compliance with free care for children under five in hospitals.

This is the only way to improve the modern fight against malaria in the Centre-Cameroon region. But one urgent action is needed: the free mass distribution of LLINs to households. With the halfway point dating back to 2016, we have to assume that many households no longer sleep under a mosquito net. According to the PNLP and PSNLP, sleeping under an impregnated mosquito net is the best way to combat malaria in the context of vector control. Unfortunately, unlike the other regions of Cameroon, the Centre region did not have an LLIN distribution campaign in 2019, and the threat of malaria was bound to persist.

CONCLUSION OF PART TWO

In the geographical area of Central Cameroon, the dynamics of the institutionalization of the fight against malaria is marked by a punctuated equilibrium and leads to ambivalent effects on the social field. Indeed, a number of tribulations punctuate the dynamics of the institutionalization of the fight against malaria in the Centre region. Whether at the level of the institutional fight or the social fight, the actors in charge of implementing the intensification of the model fight against malaria are encountering difficulties. In short, in the social and health field of Central Cameroon, the institutionalisation of the model malaria control campaign is not taking place on a 'smooth' terrain, but on a terrain strewn with pitfalls which make the power of the institutional and social actors involved in the construction and implementation of modem malaria control actions derisory. Some tribulations are directly linked to health institutions and make the fight against malaria institutionally complex: this is 'institutional complexity'. Others, on the other hand, are linked to the geographical space in which social actors are deployed and make modern public action against malaria socially complex: this is 'territorial and social complexity'.

The tribulations linked to health institutions are seriously hampering the institutional fight against malaria. This is reflected first and foremost in the entanglement of malaria control structures. This entanglement can be vertical or horizontal. The explicit corollary of this is the "overlap" between "central" and "peripheral" health institutions and the fragmented political-administrative arrangement between public health services. Secondly, the tribulations linked to health institutions are   also the result of the lack   of   positive   coordination between health facilities   .

The immediate corollary is the lack of permanent cooperation and the "complexity of joint action" between hospitals. These institutional ups and downs are a   serious obstacle to the implementation of decisions taken by central public players (the President of the Republic and MINSANTE) within health institutions.

In addition to institutional tribulations, other so-called socio-territorial tribulations influence the activities of those implementing modem malaria control actions in the social health field of Central Cameroon. These tribulations hinder the action of social actors and skew the social fight against malaria. These tribulations are directly linked to the social space in which the actors are deployed, and are referred to as 'territorial and social complexity'. Here, the tribulations result from the territory, the nationals of this territory and the 'medicamentation' that y prevails. First of all, the territory in which the social actors are deployed constitutes a serious departure from the sequences of distribution of the Mil and household enumeration, whether in the rural sub-systems or in the Yaounde sub-systems. In the rural health districts, the hostility, the extreme isolation of the area and the presence of medical deserts are all constraints on the dynamics of institutionalising the fight against malaria. In the health districts of Yaounde, it is the vastness of the territory and the predominance of 'ecologically favourable areas' that hamper the actions of social actors such as community health and enumeration agents.

Secondly, the people for whom public malaria policies are intended and the anti-malarial medication which prevails constitute serious tribulations to the hardening of the social fight against the malaria problem. In rural areas, illiteracy and poverty make the social fight against malaria more difficult. In urban areas, it is the population density and unsanitary conditions that make the fight against malaria more complex. The prescription of antimalarial drugs for these nationals also represents a serious departure from the traditional fight against malaria, as these drugs are most often sold on the informal market and are not immune to the risks of counterfeiting. In such a context, rather than being an instrument for the treatment of malaria, the drug becomes a potential danger to the health of the population and makes the process of institutionalising the fight against malaria in the geographical area of Centre-Cameroon derisory. As a result, this dynamic is producing ambivalent effects on the social field.

From the outset, the dynamic of institutionalising the fight against malaria is producing effects that are clearly visible in the social arena. First of all, the people who were being asked to change their behaviour (the target groups) have in fact changed their behaviour, which was at the root of the spread of the malaria problem, by adopting new attitudes that promote compliance with hygiene measures such as clearing drains, treating and cleaning latrines, and drinking improved water sources. These groups also scrupulously respect malaria prevention measures such as vector control and chemoprevention. In this way, LLINs/ITNs are increasingly used by households in all social categories. Pregnant women are increasingly inclined to use at least one IPT during their pregnancy.

The change in the behaviour of the target groups is leading to an improvement in the conditions of the final beneficiaries, who are increasingly seeing themselves as being free from malaria. In fact, since the *"roll back malaria" programme was adopted and implemented,* there has been a considerable drop in the malaria mortality rate in the social field of Central Cameroon, as the malaria deaths recorded in 2020 contrast sharply with those recorded in 2002. In other words, fewer and fewer people are dying of malaria in the Centre region. What's more, the number of malaria cases reported to hospitals has tended to fall over the years, as the number of patients is essentially decreasing. If we look at the curve of declared malaria cases in hospitals in Yaounde and the surrounding area between 2002 and 2020, we come to the conclusion that malaria is falling in the geographical area of Central Cameroon. As the malaria problem falls, so should the financial expenditure of households linked to this disease. This reduces the risk of poverty. So the *results of implementing* public malaria policies are clearly visible in the social arena. But this visibility of impacts and *outcomes* should not obscure the fact that the results obtained from the process of institutionalising the fight against malaria are mixed or ambivalent.

Despite the change in behaviour and the considerable drop in the number of cases of malaria recorded in hospitals in Centre-Cameroon, the problem of malaria still persists in this area of health care. Empirically, this persistence is reflected in the fact that during the year 2020, health facilities in Centre-Cameroon still recorded high numbers of malaria cases. The persistence of the malaria problem is also reflected in the fact that people in this geographical area resort to informal healthcare services and parallel pharmacies. We therefore believe that the public authorities need to pay more attention to this public health problem, by stepping up both material and symbolic actions aimed at the population. In other words, there is room for improvement in the way malaria is combated in the Centre-Cameroon region.

Conclusion generate

To sum up, we have adopted a sociological approach based on neo-institutionalism and strategic analysis for our perception of the institutionalization of the modern fight against malaria. This sociological approach would not have produced scientific and heuristic results if it had not been accompanied by data collection and processing techniques. In our view, the fight against malaria in the Centre-Cameroon region was institutionalised through a sequence of actions whose trajectory underwent bifurcation points between 1919 and 2020. Two main phases can be identified in the process of institutionalising the fight against malaria: 1919-2002 and 2002-2020. Between 1919 and 2002, the fight against malaria in the geographical area of Centre-Cameroon was marked by the privileged use of traditional methods. During the colonial period, i.e. from 1919 to 1960, the fight against malaria was characterised by the privileged intervention of non-specialist actors and by the strong presence of traditional and prophetic practices. French military doctors had the privilege of intervening in the field of malaria between 1919 and 1950, and implemented what was known as "offensive itinerant" medicine, in both prevention and treatment. Physical force was used to force local populations to receive modern health care. But with the advent of the emancipation movements in the 1950s and the exponential growth of the local population, French military medicine was put on the back burner, to be replaced by traditional and prophetic medicine. At this point, the fight against malaria experienced its first fork in the road since 1919. Local populations were now using local therapies and traditional rites to combat malaria. For the indigenous people, decolonising Cameroon meant decolonising its health system. Traditional and prophetic practices replaced military medicine, which was run by non-specialists, until 1960, when the fight against malaria was slightly modernised.

Between 1960 and 2002, the fight against malaria was slightly modernised as a result of the emergence of health professionals and the relative rationalisation of traditional medicine. The professionalisation of the fight against malaria is reflected in the "managerial reforms" of the health system in Centre-Cameroon. These reforms led to the creation of reference hospitals, a large-scale medical school and health districts. Hospitals and doctors became the major players in the institutional system for combating malaria, and the lay people who had previously been involved found themselves sidelined. The result is a situation of institutional change that Kathleen THELEN illustrates using the concept of 'institutional *layering*[509].

Moreover, as part of this process of slight modernisation in the fight against malaria, traditional medicine, once regarded as an archaic method, resurfaced and became relatively rationalised around 1990. It became "legal" with the ratification of the Alma Ata Declaration by the State of Cameroon, and with the issuing of

---

[509] See THELEN (K), "Comment les institutions evoluent perspectives d'analyse comparative historique", Op. cit... p. 30.

provisional certificates for the practice of traditional medicine to certain herbalists. This mechanism of reproduction and readaptation of previous institutions in the policies of the present corresponds to what Paul PIERSON calls 'path *dependence*'. From this point of view, the processes of institutional development in the fight against malaria initiated in 1960 in the Centre-Cameroon region are subject to the constraints imposed by the regions chosen earlier. Hence the "institutional inertia" that perpetuates traditional medicine despite the dynamics of institutional change and modernisation in the fight against malaria. Consequently, rather than being complete, the modernisation of the fight against malaria that began between 1960 and 2002 is relative, slight and limited. We had to wait until 2002 to see a real change in the dynamics of the institutionalization of modern malaria control. In fact, with the adoption of the "*Roll Back* Malaria" programme in 2002, the modern fight against the disease was stepped up, putting a definitive end to traditional malaria control methods.

In fact, from 2002 onwards, the trajectory of the fight against malaria in the geographical area of Centre-Cameroon underwent a sharp bifurcation due to both endogenous and exogenous changes in this sub-system. On an endogenous level, it is the involvement of new players in the field of malaria that is driving the dynamics of change. Exogenously, it was the adoption by the United Nations of the Millennium Development Goals (MDGs)[510] and the approval by the Cameroonian authorities of the *"Roll Back Malaria"* programme[511] that gave new impetus to the fight against malaria. In concrete terms, from 2002 onwards, the modern fight against malaria became tougher, and henceforth obeyed a multi-actor dynamic in view of the adoption of the principle of "multisectoriality". The traditional methods that had been used before the implementation of this principle gave way to modem methods. The result is a major institutional change. This change can be brought about by a wide range of actors, at different and asynchronous times, and by different social processes, also asynchronous, affecting different dimensions or aspects of the institutions or institutionalisation processes in different ways[512] . The fight against malaria is undergoing an *institutional conversion,* i.e. the new players involved in the field of malaria since 2002 are reorienting the fight against this disease towards new mandates, new objectives, new challenges and new functions. The fight against malaria is now based on new administrative methods, practices and techniques. These are being put into practice in the social field by stepping up communication to change people's behaviour, increasing the power of the public authorities and distributing mosquito nets free of charge. A new *design* and new

---

[510] A great deal of attention is devoted to the fight against malaria in Africa.

[511] This is the "Roll Back Malaria" programme, which aims to eradicate malaria in the short term. It was set up in 1998 at the instigation of Dr Gro Harlem Brundland, and is a consortium comprising the World Bank, UNDP, UNICEF and WHO. It was approved in Cameroon in 2000 and its committee began operating effectively on 29 July 2002 following decision N°0334/MPS/CAB. For more details see MOULIOM MOUNGBAKOU (LB), Op. cit... pp. 137-157.

[512] See DEMAILLY (L), GIULIANI (F) and MAROY (C), "Le changement institutionnel: processus et acteurs", Op. cit... p. 3.

cognitive scripts are thus associated with the fight against malaria.

The actors driving change and the intensification of the fight against malaria in the geographical area of Central Cameroon are both individual and collective. They are mobilising public action instruments and resources to give new meaning to the fight against malaria and to take significant action on the social front. Thus, the vast household awareness campaign as part of the CCC, public decision-making and the free distribution of LLINs are the new malaria control "policies" implemented in the Centre region and encouraged by the WHO and the PSNLP. Throughout these processes, the state retains a monopoly and is the central and dominant player.

It should be pointed out, however, that the process of institutionalising the fight against malaria, which began in 1919 during the colonial period and was reinforced in 2002, has undergone a number of setbacks. In fact, several constraints punctuate this process in the Centre region, thus hindering the equilibrium of the dynamics of change and leading to periods of major disruption. As a result, the process of institutionalising malaria control has had ambivalent effects in this geographical area.

In fact, in the geographical area of Centre-Cameroon, the dynamics of the institutionalization of the model malaria control is marked by a punctuated equilibrium and leads to ambivalent effects on the social field. In concrete terms, a certain number of tribulations punctuate the dynamics of the institutionalization of the model malaria control in the Centre region. Whether at the level of the institutional fight or the social fight, the actors in charge of implementing the intensification of the model fight against malaria are encountering difficulties. In short, in the social health field of Central Cameroon, the institutionalisation of the model malaria control campaign is not taking place on a 'smooth' terrain, but on a terrain strewn with pitfalls that make the power of the institutional and social actors involved in the construction and implementation of modem malaria control actions derisory. Some tribulations are directly linked to health institutions and make the fight against malaria institutionally complex: this is 'institutional complexity'. Others, on the other hand, are linked to the geographical space in which social actors are deployed and make modern public action against malaria socially complex: this is 'territorial and social complexity'.

The tribulations linked to health institutions are seriously hampering the institutional fight against malaria. This is reflected first and foremost in the entanglement of malaria control structures. This entanglement can be vertical or horizontal. The explicit corollary of this is the "overlap" between "central" and "peripheral" health institutions and the fragmented political-administrative arrangement between public health services. Secondly, the tribulations associated with health institutions also result from the lack of positive coordination between health facilities. The immediate corollary of this is the lack of permanent cooperation and the "complexity of joint action" between hospitals. These institutional tribulations are a serious obstacle to the implementation of decisions taken by central public players (the President of the Republic and MINSANTE) within health institutions.

In addition to the institutional tribulations, other so-called socio-territorial tribulations influence the activity of those implementing modem malaria control actions in the social health field of Central Cameroon. These tribulations hinder the action of social actors and bias the social fight against malaria. These tribulations are directly linked to the social space in which the actors are deployed, and are referred to as 'territorial and social complexity'. Here, the tribulations result from the territory, the nationals of this territory and the 'medicamentation' that y prevails. First of all, the territory in which the social actors are deployed constitutes a serious departure from the sequences of distribution of the Mil and household enumeration, whether in the rural sub-systems or in the Yaounde sub-systems. In the rural health districts, the hostility, the extreme isolation of the area and the presence of medical deserts are all constraints on the dynamics of institutionalising modern malaria control. In the health districts of Yaounde, it is the vastness of the territory and the predominance of "ecologically favourable areas" that hinder the actions of social actors such as community health and enumeration agents.

Secondly, the people for whom public malaria policies are intended and the anti-malarial medication which prevails constitute serious tribulations to the hardening of the social fight against the malaria problem. In rural areas, illiteracy and poverty make the social fight against malaria more difficult. In urban areas, it is the population density and unsanitary conditions that make the fight against malaria more complex. The prescription of antimalarial drugs for these nationals also represents a serious departure from the traditional fight against malaria, as these drugs are most often sold on the informal market and are not immune to the risks of counterfeiting. In such a context, rather than being an instrument for the treatment of malaria, the drug becomes a potential danger to the health of the population and makes the process of institutionalising the fight against malaria in the geographical area of Centre-Cameroon derisory. As a result, this dynamic is producing ambivalent effects on the social field.

The effects of the process of institutionalising the modern fight against malaria are ambivalent because, at first sight, this process produces impacts that are clearly visible in the social arena. First of all, the people who were expected to change their behaviour (the target groups) did in fact change their behaviour, which was at the root of the spread of the malaria problem, by adopting new attitudes aimed at promoting compliance with hygiene measures such as clearing drains, treating and cleaning latrines, and drinking improved water sources. These groups also scrupulously respect malaria prevention measures such as vector control and chemoprevention. In this way, LLINs/ITNs are increasingly used by households in all social categories. Pregnant women are increasingly inclined to use at least one IPT during their pregnancy.

The change in the behaviour of the target groups is leading to an improvement in the conditions of the final beneficiaries, who are increasingly seeing themselves as being free from malaria. In fact, since the *"roll back malaria" programme was* adopted and implemented, there has been a considerable drop in the malaria mortality rate in the social field of Central Cameroon, as the malaria deaths recorded in 2020 contrast sharply with those recorded in 2002. In other words, fewer and fewer people are dying of malaria in the Centre region. What's more, the number of malaria cases reported to hospitals has tended to fall over the years, as the number of patients is essentially decreasing. If we look at the curve of declared malaria cases in hospitals in Yaounde and the surrounding area between 2002 and 2020, we come to the conclusion that malaria is falling in the geographical area of Central Cameroon. As the malaria problem falls, so should the financial expenditure of households linked to this disease. This reduces the risk of poverty. So the *results of implementing* public malaria policies are clearly visible in the social arena. But this visibility of impacts and *outcomes* should not obscure the fact that the results obtained from the process of institutionalising the fight against malaria are mixed or ambivalent.

In spite of the change in people's behaviour and the considerable drop in the number of malaria cases recorded in hospitals in Centre-Cameroon, it has to be said that the malaria problem still persists in this area of health care. Empirically, this persistence is reflected in the fact that during the year 2020, health facilities in Centre-Cameroon still recorded high numbers of malaria cases. The persistence of the malaria problem is also reflected in the recourse of the populations of this geographical area to informal healthcare services and parallel pharmacies. We therefore believe that the public authorities need to pay more attention to this public health problem, by stepping up both material and symbolic actions aimed at the population.

Although the problem of malaria is still very much in evidence in the geographical area of Centre-Cameroon, despite the dynamics of modernisation, it has to be said that this so-called "modern" form of malaria control can be improved. To achieve this, the public authorities need to implement a number of measures both in the city of Yaounde and in rural areas. In the city of Yaounde, the public authorities must:

- cleaning up the city by combating ecologically unfavourable areas and insalubrious conditions;
- organise additional and regular campaigns to distribute LLINs;
- combat the sale of street drugs and counterfeit medicines by "tracking down" their promoters.

In rural areas, the fight against malaria can be improved if the public authorities :

- combat isolation;
- build modem hospitals;
- to combat illiteracy;
- ensure strict compliance with free care for children under five in hospitals.

This is the only way to improve the fight against malaria in the Centre-Cameroon region. But one urgent action is needed: free mass distribution of LLINs to households. With the halfway point dating back to 2016, we have to assume that many households are no longer sleeping under a mosquito net. According to the PNLP and PSNLP, sleeping under an impregnated mosquito net is the best way to combat malaria in the context of vector control. Unfortunately, unlike the other regions of Cameroon, the Centre region did not have an LLIN distribution campaign in 2019, and the threat of malaria was bound to persist.

I: WORKS

AMIEL (Philippe), *Ethnomethodologie appliquee: elements de sociologie praxeologique,* Presses du Lema, pp. 209. 2010.

ANGERS (Maurice), *Initiation pratique a la methodologie des sciences humaines,* Montreal, Centre Educatif et Culture, 1996 (1992).

BALANDIER (Georges), *Anthropologie politique,*Paris, PUF, 1967.

BALLE (Francis), *Medias et societe,* Paris, Montchrestien, 9th ed. 1999.

BAUMGARTNER (Frank R) and JONES (Briand D), *Agendas and Instability in American Politics,* Chicago, The University of Chicago Press, 1993.

BAYART (Jean-Francois), *L'Etat en Afrique. La politique du ventre,* Paris, Fayart, 1989.

BERGER (Peter) and LUCKMMAN (Thomas), *The Social Construction of Reality* (1966), reed. Paris, A. Colin, 1996.

BINRBAUM (Pierre) and BADIE (Bertrand), *Sociologie de 1 Etat,* Paris, Grasset, 1979.

BIRNBAUM (Pierre), *Les sommets de 1 Etat. Essai sur l'elite du pouvoir en France,* Paris, Le Seuil, 1977.

BOUDON (Raymond), *Les methodes en sociologie,* Paris, PUF, 1973.

BOURDIEU (Pierre), *Le sens pratique,* Les editions de Minuit, 1980.

BOURDIEU (Pierre), *Questions de sociologie,* Paris, Minuit, 1984.

BOURDIEU (Pierre), PASSERON (Jean-Claude) and CHAMBOREDON (Jean-Claude), *Le metier de sociologue,* Paris, Mouton, 1970.

BRAUD (Philippe), *Sociologie politique,* Paris, LGDJ, 7^e edition, 2004.

CALLON (Michel), LASCOUMES (Pierre) and BARTHE (Yannick), *Agir dans un monde incertain : essai sur la democratic technique,* Paris, editions du Seuil, 2001.

CHEVALLIER (Jacques), *Institutionspolitiques,* Paris, LGDJ, 1996

COMBESSIE (Jean-Claude), *La methode en sociologie,* 5^e edition, Paris, La Decouverte, 2007.

COULON (Alain), *L'ethnomethodologie,* Paris, PUF, (Coll. "Que sais- je ? ", Premiere edition 1987), 1996.

CROZIER (Michel) and FRIEDBERG (Erhard), *L'acteur et le systeme. Lescontraintes de 1 action collective,* Paris, Le Seuil, 1977.

CURCUFF (Philippe), *Les nouvelles sociologies,* Paris, Armand Colin, 3e ed, 2011.

DAHL (Robert), *On Democracy,* Yale University, 1998, trans. 2001.

DALOZ (Jean-Pascal), *La représentation politique,* Armand Colin, 2017.

DEPELTEAU (Francois), *La demarche d'une recherche en sciences humaines. De la question de depart a la communication des resultats,* Brussels, De Boeck, 2011 (2005).

DOBRY (Michel), *Sociologie des crises politiques: la dynamique des* *mobilisations multisectorielles,* PFNP, Paris, 2009.

DURKHEIM (Emile), *Les regies de la methode sociologique,* Paris, PUF ("Quadrige"), 1968.

EUN (Jaeho), *Sida et action publique. Une analyse du changement de politiques en France,* Paris, L'Harmattan, 2009.

FAURE (Alain), *La question territoriale. Pouvoir locaux, action publique et politique(s),* Political Science, Universite Pierre Mendes-France-Grenoble II, 2002.

FAURE (Alain) et NEGRIER (Emmanuel) (dir.), *Les politiques publiques a l'epreuve de l'action locale. Critiques de la territorialisation,* Paris, L'Harmattan, 2007.

FAVRE (Pierre) (ed.), *Sida et politique. Les premiers affrontements. 1981-1987,* Paris, L'Harmattan, 1992.

FESTINGER (Leon) and KATZ (Daniel), *Les methodes de recherche dans les sciences sociales,* PUF, Paris, 1974.

FORMEL (Michel de), OGIEN (Ruwen), QUERE (Louis) (dir.), *L'ethnomethodologie: une sociologie radicale,* Paris, La Decouverte, 2001.

FRIEDBERG (Erhard), *Le pouvoir et la regie. Dynamiques de l'action organisee,* Paris, Seuil, 1993.

GOFFMAN (Erving), *Les cadres de l'experience,* Paris, minuit, 1974.

GAUTHIER (Benoit), (dir.) *Recherche sociale. De la problematique a la collecte des données,* Ste-Foy, Presses de I'universite du Quebec, 1984.

GENIEYS (William), *Sociologie politique des elites,* Paris, Armand Colin, 2011.

GERSTLE (Jacques), *La communication politique,* 2e edition, Paris, Armand Colin, 2012.

GERSTLE (Jacques) et PIAR (Christophe), *La communication politique,* 3e edition, Paris, Armand Colin, 2016.

GINGRAS (Anne-Marie) (ed.), *La communication politique. Etat des Savoirs, enjeux et perspectives,* Presses de I'Universite du Quebec, 2003.

GINGRAS (Anne-Marie), *Medias et democratic. Le grand malentendu,* 2^e ed, Presses de I'Universite du Quebec, 2006.

GRAWITZ (Madeleine), *Methodes des sciences sociales,* Dalloz, Paris, PUF, 1996.

GRAWITZ (Madeleine) and LECA (Jean), *Trade de science politique,* Paris, PUF, Tome 3, 1985.

GRAWITZ (Madeleine) and LECA (Jean), *Trade de science politique,* Paris, PUF, Tome 4, 1985.

GUSFIELD (Joseph. R), *The culture of public problems: Drinkingdriving and symbolic order,* Chicago, University of Chicago press, 1981.

HABERMAS (Jurgen), *L'espace public. Archeologie de la publicite comme dimension constitutive de la societe bourgeoise,* Paris, Payot, 1978.

HALPERN (Charlotte), LASCOUMES (Pierre) et LE GALES (Patrick) (dir.), *L'instrumentation de 1 action publique. Controverses, resistances, effets,*

SciencePo.Les Presses, Presses de la fondation nationale de sciences politiques, 2014.

HASSENTEUFEL (Patrick), *Sociologie politique: I 'action publique, I^e* edition, Paris, Armand Colin, 2011.

HOGWOOD (Briand W.) and GUNN (Lewis A.), *Policy analysis for the real world,* Oxford, New-York, Oxford University Press, 1984.

HOOD (Christopher), *Tools of Government,* London, Macmillan, 1980.

JAMOUS (Haroun), *Sociologie de la decision. La reforme des etudes medicales et des structures hospitalieres,* Paris, CNRS, 1969.

JOBERT (Bruno) and MULLER (Pierre), *L'Etat en interaction : politiques publiques et corporatismes,* Paris, PUF, 1987.

JONES (Charles O), *An Introduction to the Study of Public Policy,* Belmont, Duxbury Press, 1970.

KATZ (Elihu) and LAZARDSFELD (Paul), *Influence personnel^,* Paris, Armand Colin, Trad. CEFAI Daniel, 2008.

KNOEPFEL (Peter), LARRUE (Corine), VARONE (Frederic) and HILL (Michael), *Public Policy Analysis,* British Library, 2007.

KNOEPFEL (Peter), LARRUE (Corine), VARONE (Frederic) and SAVARD (Jean-Francois), *Analyse et pilotage des politiques publiques,* Presses de I'Universite du Quebec, 2011.

KUBLER (Daniel) and DE MAILLARD (Jacques), *Analyser les politiques publiques,* Grenoble, PUG, 2009.

LAGROYE (Jacques) (ed.), *La politisation,* Paris, Belin, 2003.

LASCOUMES (Pierre), and LEGALES (Patrick), (dir.), *Gouvemer par les instruments,* Paris, Presses de Sciences Po, 2004.

LASCOUMES (Pierre) and LEGALES (Patrick), *Sociologie de 1 action publique,* Armand Colin, 2^e ed. 2012.

LATOUCHE (Daniel) et BEAUD (Michel), *L 'ciri de la these : comment preparer et rediger une these, un memoire ou tout autre travail universitaire,* Montreal, Boreal, 1988.

LE BART (Christian) et LEFEBRE (Remi) (dir.), *La proximite en politique. Usages, rhetoriques et pratiques,* Presses universitaires de Rennes, Rennes, 2005. 305p.

LIPSKY (Michael), *Street-Level Bureacracy: Dilemmas of the Individual in Public Services,* New York, Russell Sage Foundation, 1980.

MAC LUHAN (Marshall), *Pour comprendre les medias,* 1964, Marshall MACLUHAN, Trad. Fr. 1968, reed. Seuil, Coll. "Points essais", 1997.

MASSARDIER (Gilles), *Politiques et actions publiques,* Paris, A. Colin, 2003.

MENY (Yves) and THOENIG (Jean-Claude), *Politiques publiques,* Paris, Montchrestien, 1989.

MULLER (Pierre), *Les politiques publiques,* Paris, PUF, 1990.

MULLER (Pierre) and MENY (Yves), *L'analyse des politiques publiques,* Paris,

Montchrestion, 1998

N'DA (Paul), *Recherche et methodologie en sciences sociales et humaines. Reussir sa these, son memoire de master ou professionnel et son article,* Paris, L'Harmattan, 2015.

OLIVIER (Lawrence), BEDARD (Guy) et FERRON (Julie), *L'elaboration d'une problematique de recherche. Sources, outils et methode.* L'Harmattan, Logiques sociales, 2005.

OLSON (Mancur), *The Logic of Collective Action,* Paris, PUF, 1987.

PADIOLEAU (Jean-Gustave), *L'Etat au concret,* Paris, PUF, 1982.

PAPADOPOULOS (Yannis) *Complexite sociale et politiques publiques,* Paris, Montchrestien, 1995.

PAQUOT (Thierry), *L'espace public,* Paris, La decouverte, 2009.

PAQUIN (Stephane), BERNIER (Luc) and LACHAPELLE (Guy), (eds.), *L'analyse des politiques publiques,* PUM, 2010.

PERRET (Bernard), *L'evaluation des politiques publiques,* Paris, La Decouverte, coll "Reperes", 2001.

POMEL (Simon), (ed.) *Du risque en Afrique : terrain et perspectives,* Karthala, MSHA, 2015.

QUIVY (Raymond) and CAMPENHOUDT (Luc Van), *Manuel de recherche en sciences sociales,* Paris, Bordas/Dunod, 1988.

REZSOHAZY (Rudolf), *Pour comprendre 1 action et le changement politiques,* Louvain-la-Neuve, Duculot, 1996.

ROE (Emery M.), *Narrative Policy Analysis,* Durham, Duke University Press, 1994,

SABATIER (Paul) and JENKINS-SMITH (HANK C.), eds, *Policy Change and Learning. An Advocacy Coalition Approach,* Boulder (Co) West view Press, 1993.

SCHARPF (Fritz. W), *Games Real Actors Play. Actor-Centered Institutionalism in Policy Research,* Boulder, West view Press, 1997.

SMELSER (Neil J.) and BALTES (Paul B.), (eds), *International Encyclopedia of Social and Behavioral Sciences: Political Science,* New- York : Elsevier science, 2001.

SCHUTZ (Alfred), *Le chercheur et le quotidien* (1971), Trad. reed. Paris, Klincksieck, 1987.

VARONE (Frederic), *Le choix des instruments des politiques publiques,* Bern, Houpt, 1998.

WEBER (Max), *Economie et societe* (1922), Paris, Pion, 1991.

WEBER (Max), *Le savant et le politique* (1919), Paris, Pion, 1959.

ZAHARIADIS (Nikolaos) (ed.), *Handbook of Public Policy: agenda setting,* Cheltenham, UK-Northampton, MA, USA, Edward Elgar, 2016.

ZITTOUN (Philippe), *The political process of policymaking. A pragmatic approach to the public policy,* Palgrave, Macmillan, 2014.

II: ARTICLES AND BOOK CHAPTERS

BARTHE (Yannick), "Le recours au politique ou la problematisation politique par defaut", in LAGROYE (Jacques) (dir.), La politisation, Paris, Belin, 2003, pp. 475-492

BARTOLI (Jean-Raphael) et MERIAUX (Olivier), "Les politiques d'emploi au risque de la territorialisation concurrentielle", in FAURE (Alain) et NEGRIER (Emmanuel) (dir.), Les politiques publiques a l'epreuve de l'action locale. Critiques de la territorialisation, Paris, L'Harmattan, 2007, pp. 35-41.

BELAND (Daniel), "Neo-institutionalisme historique et politiques sociales : une perspective sociologique", in Politiques et Societes, vol. 21, 2002, pp. 21-39.

BERGERON (Henri), SUREL (Yves) et VALLUY (Jerome), "L'Advocacy Coalition Framework: une contribution au renouvellement des etudes de politique publique", in Politix, Vol 11, n°41, First quarter 1998. Political Science in the United States. Domaines et actualites, pp. 195-223.

BERNIER (Luc), "La mise en rnuvre des politiques publiques", in PAQUIN (Stephane), BERNIER (Luc) and LACHAPELLE (Guy), (eds.), L'analyse des politiques publiques, PUM, 2010, pp. 255-277.

BERTHET (Thierry), "L'Etat social a l'epreuve de l'action territoriale. Postmodemite et politiques publiques de proximite dans le champ de la relation formation-emploi", in FAURE (Alain) et NEGRIER (Emmanuel) (dir.), Les politiques publiques a l'epreuve de l'action locale. Critiques de la territorialisation, Paris, L'Harmattan, 2007, pp. 43-51.

BIAREZ (Sophie), "sphere locale et espace public", in Lien Social et politique (39), pp. 132-133.

BIRNBAUM (Pierre), "L'action de 1 Etat: differenciation et dedifferenciation", in GRAWITZ (Madeleine) et LECA (Jean), (dir.), Traite de science politique, Vol 3, Paris, PUF, 1985, pp. 643-682.

BONGRAND (Philippe) et LABORIER (Pascale), " L'entretien dans l'analyse des politiques publiques: un impense methodologique?", in Revue Fran^aise de Science Politique, Vol. 55, n° l, Fevrier 2005, pp. 73111.

BONNAL (Liliane), FAYARD (Pascal) and NANA TOMEN (Harcel), "Protection et risque maladie: le cas du paludisme au Cameroun", in Assurances et gestion des risques, Vol 82 (1-2), March-June 2015, pp. 125.

BOURDIEU (Pierre), "Decrire et prescrire : notes sur les conditions de possibilite et les limites de l'efficacites politique", in Actes de la recherche en Sciences sociales, Vol, 38, May 1981, pp. 71-73.

CHENIER (Jacques), "La specification de la problematique", in GAUTHIER (Benoit), (dir.) Recherche sociale. De la problematique a la collecte des données, Ste-Foy, Presses de l'universite du Quebec, 1984, pp. 51-77.

CHOUMIL (Elsa), "Les ambitions d'un medecin colonial en Afrique: l'histoire du docteur Jean Joseph David", in Classe Internationale, 8 fevrier 2018.

CARNEVAL (Pierre) and MOUCHET (Jean), "Vector control in Cameroon.

Passe-present-avenir. Reflexions". Manuscript n°2181, "Entomologie medicale", 28 August 2000, pp. 202-209.

COUSTEIX (Pierre-Jean), "L'art et la pharmacopee des guensseurs Ewondo (Region de Yaounde)", in Recherches et Etudes Camerounaises, 1961, pp. 1-86.

DALGALARRONDO (Sebastien) and URFALINO (Philippe), "Choix tragique, controverse et decision publique: le cas du tirage au sort des malades du sida", Revue Fran^aise de sociologie, Vol. 41, Sida et action publique, 2000, pp. 119-157.

DEBUSSMANN (Robert), "Medicalisation et pluralisme au Cameroun allemand: autorite medicale et stratégies profanes", in: outre-mer, tome 90, n°338-339, 1er semestre 2003. L'Etat et les pratiques administratives en situation coloniale, pp. 1-16.

DEMAILLY (Lise), GIULIANI (Frederique) and MAROY (Christian), "Le changement institutionnel: processus et acteurs", in Sociologies, Dossiers, Le changement institutionnel, 2019, pp. 1-16.

DI MAGGIO (Paul J) and POWELL WALTER (W), "Le neo-institutionnalisme dans 1 analyse des organisations", in Politix, vol. 10, n°40, Les sciences du politique aux Etats-Unis. Histoire et paradigmes, 1997, pp. 113-154.

DURAN (Patrice), "Le savant et la politique. Pour une approche raisonnee de 1'analyse des politiques publiques", L'annee sociologique, Vol. 40, pp. 227-259.

EBOKO (Fred), "Introduction a la question du Sida en Afrique. Politiques publiques et dynamiques sociales", in KEROUEDAN (Dominique) and EBOKO (Fred), Politiques publiques du sida en Afrique, Travaux et documents n° 61-62, CEAN, IEP Bordeaux, 1999, pp. 20-40.

EBOKO (Fred), "L'organisation de la lutte contre le sida au Cameroun : de la verticalite a la dispersion?", Bulletin de 1'APAD, 21, 2001, pp. 119.

EBOKO (Fred), "Politique publique et sida en Afrique", in cahiers d'etudes africaines, n°178, 2005, pp. 1-31.

EYMERI (Jean-Michel), "Frontiere ou marches? De la contribution de la haute administration a la production du politique" in LAGROYE (Jacques) (ed.), La politisation, Paris, Belin, 2003, pp. 47-77.

FEYT (Gregoire), "Redistribution des pouvoirs, redistribution des cartes. La connaissance des territoires, enjeu inedit de 1'action publique?" in FAURE (Alain) et NEGRIER (Emmanuel) (dir.), Les politiques publiques a I'epreuve de 1'action locale. Critiques de la territorialisation, Paris, L'Harmattan, 2007, pp. 107-115.

GARRAUD (Philippe), "Agenda/Emergence", in BOUSSAGUET (Laurie), JACQUOT (Sophie) and RAVINET (Pauline), Dictionnaire des politiques publiques, 2e edition, Presses de la fondation nationale des Sciences Politiques, 2006, pp. 51-59.

GARRAUD (Philippe), "Politiques nationales : elaboration d'un agenda", I'annee sociologique, n°40, pp. 17-41.

GAXIE (Daniel), "Une construction mediatique du spectacle politique? Realite et limites de la contribution des medias au developpement des perceptions negatives du politique", in LAGROYE (Jacques) (ed.), La politisation, Paris, Belin, 2003, pp. 325-356.

GEEST (Sjaak Van Der), "Les medicaments sur un marché Camerounais. Reconsideration de la commodification et de la pharmaceuticalisation de la sante", in revue Internationale francophone d'anthropologie de la sante, Association Amades, Varia, 2017, pp. 1-15.

GROSSMAN (Emiliano), "Acteur", in BOUSSAGUET (Laurie), JACQUOT (Sophie) and RAVINET (Pauline), Dictionnaire des politiques publiques, 2^e edition, Presses de la fondation nationale des Sciences Politiques, 2006, pp. 25-31.

HALL (Peter A) and TAYLOR (Rosemary C.R), "La science politique et les trois neo-institutionnalismes", in Revue Fran^aise de Science Politique, 47^e annee, n°3-4, 1997, pp. 469-496.

HAMIDI (Camille), "Elements pour une approche mteractionniste de la politisation: engagement associatif et rapport au politique dans des associations locales issues de 1 immigration", in Revue Fran^aise des Sciences Politiques, 2006, Vol. 56 (№1), pp. 5-25.

HALPERN (Charlotte), LASCOUMES (Pierre) ET LE GALES (Patrick), " L'instrumentation et ses effets. Debats et mises en perspective theorique", in HALPERN (Charlotte), LASCOUMES (Pierre) et LE GALES (Patrick) (dir.), L'instrumentation de l'action publique. Controverses, resistances, effets, SciencePo.Les Presses, Presses de la fondation nationale de sciences politiques, 2014, pp. 15-59.

HASSENTEUFEL (Patrick), "L'Etat mis a nu par les politiques publiques", in BADIE (Bertrand) et DELOYE (Yves) (dir.), Le temps de 1 Etat, melange en l'honneur de Pierre BIRNBAUM, Fayard, 2007, pp. 311-329.

JACQUOT (Sophie), "L'approche sequentielle", in BOUSSAGUET (Laurie), JACQUOT (Sophie) and RAVINET (Pauline), Dictionnaire des politiques publiques, 2^e edition, Presses de la fondation nationale des Sciences Politiques, 2006, pp. 73-80.

JOBERT (Bruno), "L'Etat en action. L'apport des politiques publiques", in Revue Fran^aise de Science Politique, 35^e annee, n° 4, 1985, pp. 654682.

LAGROYE (Jacques), "On ne subit pas son role", Politix n° 38, 1997, pp. 7-17.

LASCOUMES (Pierre), "Gouvemer par les instruments. Ou comment s'instrumente l'action publique", in LAGROYE (Jacques) (ed.), La politisation, Paris, Belin, 2003, pp. 387-401.

LASCOUMES (Pierre) and SIMARD (Louis), "L'action publique au prisme de ses instruments", Revue Fran^aise de Science Politique, 2011, Vol. 61, pp. 5-22.

LECOURS (Andre), "L'approche neo-institutionnaliste en science politique : unite ou diversite?" in Politique et Societe, Vol. 21, 2002, pp. 3-19.

LEVY (Julien) and WARIN (Philippe), "Ressortissants", in BOUSSAGUET (Laurie), JACQUOT (Sophie) and RAVINET (Pauline), Dictionnaire des politiques publiques, 5^e edition, Presses de la fondation nationale des Sciences Politiques,

2019, pp 555-561.

LINDBLOM (Charles), "The science of 'muddling through'", Public Adminitration Review, 19 (2), 1959, pp. 79-88.

MASSE (Jolicoeur M), "Introduction au modele de l'equilibre ponctue : un modele pour comprendre la stabilite et les changements radicaux en politiques publiques", Montreal, Quebec: National Collaborating Centre for Healthy Public Policy, 2018, pp. 1-11.

MANFET (Christelle), "La territorialisation problematique de l'action universitaire. Dynamique sectorielle et reponses locales", in FAURE (Alain) et NEGRIER (Emmanuel) (dir.), Les politiques publiques a l'epreuve de l'action locale. Critiques de la territorialisation, Paris, L'Harmattan, 2007, pp. 27-34.

MEGIE (Antoine), "Mise en rnuvre", in BOUSSAGUET (Laurie), JACQUOT (Sophie) and RAVINET (Pauline), Dictionnaire des politiques publiques, 2^e edition, Presses de la fondation nationale des Sciences Politiques, 2006, pp. 285-292.

MEVA'A ABOMO (Dominique), "Le fardeau de la lutte centre le paludisme urbain au Cameroun: etat des lieux, contraintes et perspectives", RCGT, Vol 3 (2), 2016, pp. 26-42.

MOULIOUM MOUNGBAKOU (Ibrahim Bienvenu), "Concurrence des therapeutiques traditionnelles et biomedicales dans la lutte centre le paludisme a I extreme-nord du Cameroun", in Journal des anthropologues, 2014, pp. 137-157.

MULLER (Pierre), "Esquisse d'une theone du changement dans 1 action publique". Structures, acteurs et cadres cognitifs", in Revue Fran^aise de Science politique, 2005/1 Vol. 55, pp. 155-187.

MULLER (Pierre), "L'analyse cognitive des politiques publiques : vers une sociologie politique de l'action publique", in Revue Fran^aise de Science Politique, 50^e annee, n°2, 2000, pp. 189-208.

MULLER (Pierre), "Referentiel", in BOUSSAGUET (Laurie), JACQUOT (Sophie) and RAVINET (Pauline), Dictionnaire des politiques publiques, 2^e edition, Presses de la fondation nationale des Sciences Politiques, 2006, pp. 372-378.

NAHRATH (Stephane), "Les referentiels de politiques publiques", Congres annuels de 1 association suisse de Science Politique, Universite de Geneve, 2010, pp. 1-25.

NAY (Olivier), "La politique des bons offices. L'elu, l'action publique et le territoire", in LAGROYE (Jacques) (ed.), La politisation, Paris, Belin, 2003, pp. 199-219.

KONTCHOU KOUOMENGI (Augustin), "Methode de recherche et domaines nouveaux en Relation Internationale", in Revue Camerounaise des Relations intemationales, Ed. Speciale, 1982, pp. 55-60.

OKALLA (Raphael), "Le district de sante d'Akonolinga", in Bulletin de l'APAD, 2001, pp. 1-11.

OKALLA (Raphael) and LE VIGOUROUX (Alain), "Cameroun: de la reorientation des soins de sante primaries au plan de développement sanitaire", in Bulletin de 1 APAD, 2001, pp. 1-10.

OGIEN (Albert), "Garfinkel et la naissance de I'ethnomethodologie", Paris, Institut marcel Mauss-CEMS, April 2016, pp. 1-16.

PALIER (Bruno) and SUREL (Y), "Les trois 'i' et l'analyse de 1 Etat en action", Revue Fran^aise de Science Politique, 2005/1, Vol.55, pp. 7-32.

PAQUOT (Thierry), "Qu'est-ce qu'un territoire?", in Vie Sociale, n°2, 2011, pp. 23-32.

PIERSON (Paul), "Path Dependence, Increasing Returns, and the Study of Politics", American Political Science Review, 94 (2), 2000, pp. 251-267.

PROULX (Serge) et BELANGER (Danielle), "La reception des messages", in GINGRAS (Anne-Marie) (dir.), La communication politique. Etat des Savoirs, enjeux et perspectives, Presses de I'Universite du Quebec, 2003, pp. 215-255.

QUERMONNE (Jean-Louis), "Les politiques institutionnelles. Essai d'interpretation et de typologie", in GRAWITZ (Madeleine) and LECA (Jean), Traite de science politique, Paris, PUF, Tome 4, 1985, pp. 61-88.

SAURUGGER (Sabine), "Groupe d'interet", in BOUSSAGUET (Laurie), JACQUOT (Sophie) and RAVINET (Pauline), Dictionnaire des politiques publiques, 2^e edition, Presses de la fondation nationale des Sciences Politiques, 2006, pp. 252-260.

SHEPPARD (Elisabeth), "Probleme public", in BOUSSAGUET (Laurie), JACQUOT (Sophie) and RAVINET (Pauline), Dictionnaire des politiques publiques, 2^e edition, Presses de la fondation nationale des Sciences Politiques, 2006, pp. 349-355.

SIMEANT (Johanna), "Un humanitaire 'apolitique' ? Demarcations, socialisations au politique et espaces de realisation de soi", in LAGROYE (Jacques) (ed.), La politisation, Paris, Belin, 2003, pp. 163-196.

SORIAT (Clement), "L'implication des acteurs associatifs beninois dans 1'action publique de lutte contre le sida: entre domestication et prise de pouvoir", in Penser 1'action publique en contexte africain, Congres AFSP Aix, 2015, pp. 1-18.

SUREL (Yves), "Approches cognitives", in BOUSSAGUET (Laurie), JACQUOT (Sophie) and RAVINET (Pauline), Dictionnaire des politiques publiques, 2^e edition, Presses de la fondation nationale des Sciences Politiques, 2006, pp. 80-88.

TCHOUPIE (Andre), "L'institutionnalisation des deliberations dans 1 espace public au sein des chefferies Bamileke de 1 Ouest-Cameroun", in Afrique et Developpement, CODESRIA, Vol. XXXIV, 2009, pp. 65-92.

THELEN (Kathleen), "Comment les institutions evoluent: perspectives d'analyse comparative historique", L'annee de la regulation, 7, 2003-2004, pp. 13-44.

THOENIG (Jean-Claude), "L'analyse des politiques publiques", in GRAWITZ (Madeleine) and LECA (Jean), Traite de science politique, Paris, PUF, Tome 4, 1985, pp. 1-60.

THOENIG (Jean-Claude), "Politique publique", in BOUSSAGUET (Laurie), JACQUOT (Sophie) and RAVINET (Pauline), Dictionnaire des politiques publiques, 2^e edition, Presses de la fondation nationale des Sciences Politiques, 2006, pp. 328-335.

VERNAZZA-LICHT (Nicole), BLEY (Daniel), MUDUDU (Leon) and MBETOUMOU (Marceline), "Entre fatalite et action: perception et gestion du risque palustre au Cameroun", in POMEL (Simon), (ed.) Du risque en Afrique: terrain et perspectives, Karthala, MSHA, 2015, pp. 202-213.

VOLLET (Dominique), CALLOIS (Jean-Marc), MOQUAY (Patrick) et ROUSSEL (Veronique), "Les politiques rurales gagnees par la territorialisation", in FAURE (Alain) et NEGRIER (Emmanuel) (dir.), Les politiques publiques a I'epreuve de 1 action locale. Critiques de la territorialisation, Paris, L'Harmattan, 2007, pp. 79-85.

WATSON (Rodney), "Continuite et transformation de I'ethnomethodologie", in FORMEL (Michel de), OGIEN (Ruwen), QUERE (Louis) (dir.), L'ethnomethodologie: une sociologie radicale, Paris, La Decouverte, 2001, pp. 17-29.

ZAHARIADIS (Nicolaos), "Setting the agenda on agenda setting: definitions, concepts and controversies", in ZAHARIADIS (Nikolaos) (ed.), Handbook of Public Policy: agenda setting, Cheltenham, UK- Northampton, MA, USA, Edward Elgar, 2016, pp. 1-21.

ZITTOUN (Philippe), "La fabrique pragmatique des politiques publiques", in Anthropologie et Developpement, n°45/2017, pp. 65-89.

III: THESES AND DISSERTATIONS

BONVALET (Perrine), *De I'urgence politique a la gestion de 1 action publique: Construire et institutionnaliser I'acces aux traitements du VIH/Sida au Benin,* These de doctorat en science politique, Universite de Bordeaux, 2014.

DJEUTCHOUANG SAYANG (Collins), *Interet de l'utilisation de tests de diagnostic rapide du paludisme sur le cout et I'efficacite de la prise en charge des patients febriles a Yaounde Cameroun,* These de doctorat en pathologie humaine, Faculte de medecine de Marseille, 2010.

DESCLAUX (Alice), *L'epidemie invisible. Anthropologie d'un systeme medical a I'epreuve du sida chez 1'enfant a Bobo Dioulasso, Burkina Faso,* These de doctorat en Anthropologie, Aix-Marseille 3, 1997.

DIAW (Mamadou), *L'appropriation communautaire des cases de sante selon la perspective des populations,* These de doctorat en Geographie, Universite Paris Saclay, 2019.

EBOKO (Fred), *Pouvoir, jeunesse et sida au Cameroun: politiques publiques, dynamiques sociales et constructions des sujets,* These de doctorat en science politique, Universite Montesquieu-Bordeaux IV, Institut d'Etudes politiques de Bordeaux, Ecole doctorale de science politique de Bordeaux, Bordeaux, 2002.

FOE NDI (Christophe), *La mise en oeuvre du droit a la sante au Cameroun,*

These de doctorat en Droit public, Universite d'Avignon, 2019.

KOJOUE KAMGA (Larissa), *Enfants et VIH/Sida au Cameroun. Construction et implications de l'agenda politique,* These de doctorat en science politique, Universite de Bordeaux Segalen, SciencePo Bordeaux, 2013.

MBOUZEKO (Raymond), *Discours et representations sociales dans la prevention du paludisme au Cameroun. Logique des discours, perceptions de la maladie et pratiques des populations,* These de doctorat en science de l'information et de la communication, Universite Lumiere de Lyon 2, 24 novembre 2010.

SIMBE AVORE (Thomas D'Aquin), *Les politiques publiques de traitement du paludisme dans l'arrondissement de Mbouda (1952-2014),* Memoire de master en science politique, Universite de Dschang, 2016.

TSOTSA (Nathanael Edrich), *L 'action publique de lutte contre le VIH/Sida. Acteurs, controverses et dynamiques. Analyse comparee a partir des exemples sud-africain, Burkinabe et camerounais,* These de doctorat en science politique, Institut d'etude politique de Bordeaux, 2009.

IV: DICTIONARIES

DICTIONNAIRE DES POLITIQUES PUBLIQUES, edited by Laurie BOUSSAGUET, Sophie JACQUOT and PAULINE RAVINET, 2e revised and corrected edition, Paris, Presses de sciences Po, 2006.

DICTIONNAIRE DES POLITIQUES PUBLIQUES, edited by Laurie BOUSSAGUET, Sophie JACQUOT and PAULINE RAVINET, 5e édition entièrementment mise a jour et augmentee, Paris, Presses de sciences Po, 2019.

DICTIONNAIRE DE LA SCIENCE POLITIQUE ET DES INSTITUTIONS POLITIQUES, edited by Guy HERMET, Bertrand BADIE, Pierre BIRNBAUM and Philippe BRAUD, 7e revised and expanded edition, Paris, Armand Colin, 2010.

DICTIONNAIRE DE SCIENCES POLITIQUES ET SOCIALES, edited by David ALCAUD and Laurent BOUVET, Edition Sirey, Paris, 2004.

DICTIONNAIRE UNIVERSEL, Hachette, Edicef, 2001.

LE DICTIONNAIRE DES SCIENCES SOCIALES, edited by Jean-Francois DORTIER, Editions sciences humaines, 2013.

LEXIQUE DE SCIENCE POLITIQUE, edited by Olivier NAY, 4^e edition, Dalloz, 2017.

V: REPORTS AND ADMINISTRATIVE DOCUMENTS

Ministerial decision n°0406/D/MINSANTE/CAB of 23 June 2014 on the pricing of the management of severe malaria in Cameroon.

Decree n°2013/093 of 03 April 2013 on the organisation of the Ministry of Public Health.

DESCLAUX (Alice), *Une etrange absence de crise. L'anthropologic et la faillite du systeme,* oral presentation at the l'AMADES conference, "L'anthropologie des systemes de sante", Paris 6-8 January 1999.

IRESCO, *Etude d'évaluation du projet de distribution des moustiquaires impregnees a l'insecticide par l'association camerounaise pour marketing social,* February 2006.

MINSANTE, *Enquete post campagne sur l utilisation des MILDA 2016/2017. Final report,* December 2017.

MINSANTE-OMS, ONSP, *Rapport de suivi des 100 indicateurs cles de sante au Cameroun en 2019, focus sur les ODD,* 2019.

Law №90-035 of 10 August 1990 on the practice and organisation of the profession of pharmacist.

Law №96/06 of 18 January 1996 revising the Constitution of 2 June 1972.

Reseau Sida Afrique, *Cartographic of the fight against malaria in Cameroon,* Kid AIDS.

Plan strategique national de lutte contre le paludisme au Cameroun 20072010, The Global Fund, To fight AIDS , Tuberculosis and malaria.

VI: ELECTRONIC DOCUMENTS AND SOURCES

BIAREZ (Sophie), "Sphere locale et espace public", <u>Lien social et politiques,</u> (39), https://doi. 0rg/10.7202/0005059ar. Accessed 11 January 2020. pp. 132-133.

Claude (G), "Data collection: characteristics, techniques and examples. Scribbr. <u>https://www.scribbr.fr/methodologie/collecte-de-donnees/</u>, 24 March, 2021.

BIRABALUGE Louis, "Apprendre d'un pretre guensseur : Pere Memrad- Pierre HEBGA", available online at <u>https://afrique.xaveriens.org/collaborateurs/item/louisbirabaluge-sx</u>

"Cameroon community immersion project. Theme: the fight against malaria in a peripheral hospital", www.medecine.unige.ch.pdf, consulted on 13 March 2019. <u>https://ioumals.openedition.org/faceaface/445?lang=en</u>

"Les acquis et les limites de 1 analyse strategique", document available online at <u>https://www.lc-doc.com</u>

www.infosdafrique.com.pdf. Accessed 18 June 2020.

www.vumpu.com.pdf. Accessed on 18 June 2020. ...., ...

<u>www.minsante.cm,</u> consulted on 23 April 2020.

VII: PRESS

- *Cameroon tribune,* 14/11/2005.
- *Cameroon tribune,* 16/11/2005.
- *Cameroon tribune,* 18/11/2005.
- *Cameroon tribune,* 08/04/2008.
- *Cameroon Tribune,* 30/08/2010.
- *Jeune Afrique,* 30/10/2013.
- *Lanouvelleexpression   ,   24/04/2001.*
- *Lanouvelleexpression   ,   04/02/2001.*
- *Lanouvelleexpression   ,   21/11/2005.*
- *Lanouvelleexpression   ,   08/06/2007.*
- *Lanouvelleexpression   ,   20/03/2008.*
- *Lanouvelleexpression   ,   01/18/2008.*

APPENDICES

APPENDIX 1
INDICATIVE INTERVIEW GUIDE
HEALTH PROFESSIONALS

1. When or which year did you start to see an upsurge in malaria cases?
2. What strategies are used by health facilities to treat patients suffering from malaria?
3. Why do some patients refuse to go to hospital if they are ill?
4. What treatment do you prescribe for patients with suspected malaria?
5. How many cases of malaria did you record in 2020?
6. Have you ever recorded any cases of malaria-related deaths in this health facility?
7. Are children aged zero to five treated free of charge for malaria in this health facility?
8. How much does a RDT cost in this health facility?
9. Do you administer IPT to pregnant women?
10. What difficulties do you encounter when treating malaria patients?
11. Why do you think malaria persists in the region?

POLITICAL AND ADMINISTRATIVE AUTHORITIES

1. What are you doing in the field to combat malaria?
2. *What is the "small no be sick" programme?*
3. Are you taking any specific measures in your locality to combat malaria? If so, what measures?
4. What difficulties are you encountering in the fight against malaria?
5. What role did you play in the various campaigns to distribute ITNs/MILDAs to households?
6. Why do you think malaria persists in the region?

COMMUNITY PLAYERS (NGOS AND ASSOCIATIONS)

1. What do you think of the *"roll back malaria"* and *"nightwatch"* programmes, and what role do you play in them?
2. What are you doing in the field to reduce malaria?
3. Were you involved in the process of distributing ITNs/MILDAs to the population during the so-called "mass" campaigns?
4. Do you treat cases of malaria?
5. Are you more successful at deploying in rural than in urban areas?
6. What difficulties do you encounter in the field?
7. Why do you think malaria still persists in the region?

These questions, which took the form of interviews, enabled us to collect data in the field. They were put to health professionals, in particular those from the Centre regional public health delegation, and those from the Biyem-Assi, Cite-Verte, Ntui and Yoko health districts. They were also put to the political and administrative authorities, in particular the former deputy for Mefou-et-Akono and the deputy prefect of Mfoundi. Finally, these questions were addressed to certain NGO leaders, in particular those of Malaria No More (MNM) and the Cameroon Association for Social Marketing (ACMS). We have taken care to quote these different people and the different dates of the interviews for the purposes of our thesis.

APPENDIX 2
LIST OF ASSOCIATIONS/NGOs FIGHTING MALARIA IN THE
CENTRE
REGION

№	NAME OF ASSOCIATION OR NGO	FIELD OF ACTIVITY
1	CAMEROUN MEDIA AGAINST MALARIA (CAMAM)	COMMUNICATION
2	CAMEROUN ASSOCIATION OF SCHOOL ADMINISTRATION	SCHOOL HEALTH
3	CATHOLIC HEALTH ORGANISATION IN CAMEROON	SERVICE PROVISION, ADVOCACY
4	ASSOCIATION DES EDUCATEURS EMERITES (ASEDEM)	COMBATING POVERTY AMONG YOUNG PEOPLE BY USING TEACHING TECHNIQUES TO PROMOTE MEASURES TO COMBAT MALARIA IN SCHOOLS
5	CAMEROON SOCIETY OF PAEDIATRICS	CHILD HEALTH
6	BESAFCA/ SOS MEDECIN	MEDICAL CARE, TRAINING OF LOCAL PLAYERS, RESEARCH AND DECENTRALISED CARE
7	CAMEROON NATIONAL ASSOCIATION FOI FAMILY WELFARE (CAMNAFAW)	REPRODUCTIVE HEALTH, SOCIAL MOBILISATION, CLINICAL SERVICE PROVISION
8	ASSOCIATION OF OBSTETRIC GYNECOLOGISTS OF CAMEROON (SOGOC)	REPRODUCTIVE HEALTH
9	ASSOCIATIONS OF DYNAMIC WOMEN IN CAMEROON	CARE FOR VULNERABLE CHILDREN IN THE MBAM-ET- INOUBOU, MEFOU AND AKONO REGIONS
10	MALARIA NO MORE	AWARENESS-RAISING,

		MONITORING AND EVALUATION, SOCIAL ACTIONS, CARE.
11	AFLUSES	RAISING AWARENESS ON THE GROUND WITHIN COMMUNITIES
12	HUMANITARIAN ASSOCIATION OF MEDICAL RESEARCHERS	TRADITIONAL MEDICINE
13	FESADE	MATERNAL AND CHILD HEALTH, THE FIGHT AGAINST MALARIA, YOUTH HEALTH
14	ROSACAM	ALL ASPECTS OF REPRODUCTIVE HEALTH, THE FIGHT AGAINST MALARIA AND TUBERCULOSIS
15	INTERNATIONAL MISSION FOR CHRIST (MIC)	ADVOCACY, RAISING PUBLIC AWARENESS OF THE FIGHT AGAINST MALARIA
16	AFRICAN SYNERGY AGAINST AIDS AND SUFFERING	THE FIGHT AGAINST HIV/AIDS AND OTHER ENDEMICS, THE FIGHT AGAINST MATERNAL INFANT MORTALITY, THE ALLEVIATION OF POVERTY
		CHILDREN'S SUFFERING
17	IMPACT SANTE AFRIQUE	ADVOCACY, AWARENESS-RAISING, MONITORING AND EVALUATION
18	ASSOCIATION CAMEROUNAISE POUR LE MARKETING SOCIAL	COMBATING HIV/AIDS, COMBATING MALARIA, COMBATING WATER-BORNE DIARRHOEAL DISEASES, CHILDCARE AND FAMILY PLANNING
19	NATIONAL FEDERATION OF TRADE AND SERVICES UNIONS OF CAMEROON (FESCOSCAM)	PREVENTION, SOCIAL MOBILISATION, COMMUNICATION, VECTOR CONTROL

| 20 | BCH AFRICA | BUILDING COMMUNITY CAPACITY IN HEALTH COMMUNICATION |
| 21 | RACTAP | MONITORING, ASSESSMENT, CARE |

APPENDIX 3
DECISION N°0406/D/MINSANTE/CAB OF 23 JUNE 2014
ON THE PRICING OF TREATMENT FOR
SEVERE PAUUDISM IN CAMEROON

<table>
<tr><td align="center">REPUBLIQUE DU CAMEROUN
Paix – Travail – Patrie
————————
MINISTERE DE LA SANTE PUBLIQUE
————————
CABINET DU MINISTRE
————————</td><td align="center">REPUBLIC OF CAMEROON
Peace – Work – Fatherland
————————
MINISTRY OF PUBLIC HEALTH
————————
MINISTER'S OFFICE
————————</td></tr>
</table>

DECISION N° 0 4 0 6 /D/MINSANTE/ CAB du _______ 2 3 JUIN 2014

PORTANT TARIFICATION DE LA PRISE EN CHARGE DU PALUDISME GRAVE AU CAMEROUN

LE MINISTRE DE LA SANTE PUBLIQUE,

Vu   la Constitution ;

Vu   la Loi n°96/03 du 04 janvier 1996 portant Loi-cadre dans le domaine de la Santé ;

Vu   la Loi n°90/62 du 19 décembre 1990 portant dérogation spéciale aux formations sanitaires publiques en matière financière ;

Vu   le Décret n°2013/093 du 03 avril 2013 portant organisation du Ministère de la Santé Publique ;

Vu   le Décret N°2011/408 du 09 décembre 2011 portant organisation du Gouvernement ;

Vu   le Décret n° 2011/410 du 09 décembre 2011 portant formation du Gouvernement ;

Vu   le Décret 2005/252 du 30 juin 2005 portant création, organisation et fonctionnement de la Centrale Nationale d'Approvisionnement en Médicaments et Consommables Médicaux Essentiels modifié et complété par le décret n°2009/386 du 30 novembre 2009 ;

Vu   le Décret n°93/228/PM du 15 mars 1993 fixant les modalités d'application de la Loi n.°90/062 du 19 décembre 1990, accordant dérogation spéciale aux formations sanitaires publiques en matière financière ;

Vu   la Décision n°0399/D/MINSANTE/CAB du 18 juin 2014 portant directives sur la gratuité du traitement du paludisme grave chez les enfants de moins de cinq ans ;

Vu   la Décision n°0092/D/MSP/CAB du 29 novembre 2001 portant tarification des médicaments et dispositifs médicaux essentiels au niveau des Centres d'Approvisionnement Pharmaceutique Régionaux et des formations sanitaires publiques ;

Vu   la Lettre Circulaire n°D36-11/LC/MINSANTE/CAB du 05 Mars 2012 portant révision des prix des produits pharmaceutiques dans le secteur public ;

Considérant les nécessités de service.

DECIDE :

Article 1er:

La présente Décision porte tarification de la prise en charge du paludisme grave par l'artésunate injectable ou l'artémether injectable.

Article 2:

(1) L'artésunate injectable et l'artémether injectable utilisés dans le cadre du traitement du paludisme grave sont cédés à titre gratuit à tous les patients quelque soit l'âge.

(2) Toutefois, les formations sanitaires s'acquittent auprès des Centres d'Approvisionnement Pharmaceutique Régionaux ou des Fonds Régionaux de Promotion de la Santé, des frais de gestion de ces antipaludiques cédés à titre gratuit.

Article 3 :

Les molécules et intrants complémentaires à la prise en charge du paludisme grave sont : un (01) antipyrétique à base de paracétamol, un (01) test de diagnostic rapide, 06 seringues de 10 ml, une (01) poche sérum glucosé 5%, une (01) poche de ringer lactate, un (01) perfuseur, trois (03) cathéters G24 ou trois (03) épicrâniennes G24.

Article 4:

(1) Ces molécules et intrants complémentaires ne sont pas gratuits ; toutefois, ils sont subventionnés à 100% chez les enfants de moins de cinq ans par un mécanisme de solidarité inter-génération.

(2) La solidarité inter génération consiste à faire supporter le coût des malades de paludisme grave âgés de moins de cinq ans par les malades de paludisme grave âgés de plus de 5 ans, sachant que ces derniers représentent 65% et les premiers 35% (Données PNLP, 2013).

Article 5:

Les molécules et intrants complémentaires à l'artésunate injectable ou l'artémether injectable sont partiellement subventionnés pour les femmes enceintes qui représentent 10% des malades, par les patients âgés de plus de cinq ans (Données PNLP, 2013).

Article 6:

(1) La molécule antipaludique du paludisme grave de la **femme enceinte au premier trimestre de la grossesse est uniquement la quinine**, tel que précisé dans le Guide National de Prise en Charge du Paludisme au Cameroun.

(2) La quinine n'est pas gratuite.

Article 7:

Les molécules et intrants rentrant dans le traitement des complications éventuelles du paludisme grave font l'objet d'une facturation quelque soit l'âge du patient.

Article 8:

(1) Les formations sanitaires acquièrent normalement aux prix homologués les molécules et intrants complémentaires auprès des Centres d'Approvisionnement Pharmaceutique Régionaux ou des Fonds Régionaux de Promotion de la Santé.

(2) Les frais de gestion des Centres d'Approvisionnement Pharmaceutique Régionaux ou des Fonds Régionaux de Promotion de la Santé sont facturés aux formations sanitaires sur la base des quantités de molécules antipaludiques destinées aux adultes à raison de **600 francs CFA par traitement** (12 ampoules d'artésunate injectable 60mg ou 6 ampoules d'artémether injectable 80mg).

Article 9:

Les prix de cession du traitement **dans le cadre de la solidarité inter-génération d'un épisode de paludisme grave** au niveau des formations sanitaires s'établissent ainsi qu'il suit :

- enfants de moins de cinq ans (hors frais d'hospitalisation et de complications éventuelles) : **gratuit** ;
- femmes enceintes à partir du deuxième trimestre de grossesse (hors frais d'hospitalisation et de complications éventuelles) : **quatre mille (4 000) francs CFA** ;
- patients de plus de cinq ans (hors frais d'hospitalisation et de complications éventuelles) : **huit mille (8 000) francs CFA.**

Article 10: La présente Décision abroge la Décision N°0243/MINSANTE/SG/DAJC/ du 30 avril 2014 et sera enregistrée, publiée puis communiquée partout où besoin sera /-

LE MINISTRE DE LA SANTE PUBLIQUE,

André MAMA FOUDA

APPENDIX 4
INFORMATION NOTE ON THE 2015/2016 NATIONAL MILDA DISTRIBUTION CAMPAIGN

<table>
<tr><td>RÉPUBLIQUE DU CAMEROUN
Paix-Travail-Patrie
MINISTÈRE DE LA SANTÉ PUBLIQUE</td><td></td><td>REPUBLIC OF CAMEROON
Peace-Work-Fatherland
MINISTRY OF PUBLIC HEALTH</td></tr>
</table>

NOTE D'INFORMATION
2ème CAMPAGNE NATIONALE DE DISTRIBUTION GRATUITE DE 12,350 MILLIONS DE MILDA

Le Cameroun s'est engagé à atteindre la « Couverture universelle des populations en Moustiquaires Imprégnées d'insecticides de Longue Durée d'Action (MILDA) », afin de diminuer d'une manière significative la morbidité et la mortalité dues au paludisme.

Une deuxième campagne nationale de distribution gratuite d'environ **12 350 000** de MILDA a débuté le 2e semestre 2015 et va s'achever en Mai 2016.

Au 1er Avril 2016, Neuf Régions sur dix ont déjà reçu **8.765.522 MILDA** réparties comme suit :

Régions	Nombre de MILDA distribuées	Précisions
Adamaoua	697 220	
Est	496 377	
Extrême-Nord	1 953 921	Données de 28 districts de santé sur 30
Littoral	1 473 212	Données de 23 districts de santé sur 24
Ouest	734 099	Données de 18 districts de santé sur 20
Nord	1 362 401	
Nord-Ouest	956 407	
Sud	376 767	
Sud-Ouest	715 118	
Total	8 765 522	

La 10ème Région à savoir le Centre, recevra les MILDA la mi- Mai 2016, à l'issue des enquêtes de dénombrement programmées la dernière semaine d'Avril 2016.

Yaoundé le 15 AVR 2016

André MAMA FOUDA
Ministre de la Santé Publique

I want morebooks!

Buy your books fast and straightforward online - at one of world's fastest growing online book stores! Environmentally sound due to Print-on-Demand technologies.

Buy your books online at
www.morebooks.shop

Kaufen Sie Ihre Bücher schnell und unkompliziert online – auf einer der am schnellsten wachsenden Buchhandelsplattformen weltweit! Dank Print-On-Demand umwelt- und ressourcenschonend produzi ert.

Bücher schneller online kaufen
www.morebooks.shop

Printed by Books on Demand GmbH, Norderstedt / Germany